CAMBRIDGE MONOGRAPHS IN
EXPERIMENTAL BIOLOGY
No. 11

EDITORS:
T. A. BENNET-CLARK, P. B. MEDAWAR
GEORGE SALT (*General Editor*)
C. H. WADDINGTON, V. B. WIGGLESWORTH

INTERMEDIARY METABOLISM
IN PLANTS

THE SERIES

INTERMEDIARY METABOLISM IN PLANTS

BY

DAVID D. DAVIES

Division of Food Preservation, C.S.I.R.O.
Sydney, Australia

CAMBRIDGE
AT THE UNIVERSITY PRESS
1961

CAMBRIDGE UNIVERSITY PRESS
Cambridge, New York, Melbourne, Madrid, Cape Town,
Singapore, São Paulo, Delhi, Tokyo, Mexico City

Cambridge University Press
The Edinburgh Building, Cambridge CB2 8RU, UK

Published in the United States of America by Cambridge University Press, New York

www.cambridge.org
Information on this title: www.cambridge.org/9780521279291

First published 1961
First paperback edition 2011

A catalogue record for this publication is available from the British Library

ISBN 978-0-521-04792-0 Hardback
ISBN 978-0-521-27929-1 Paperback

CONTENTS

ACKNOWLEDGMENTS

IT is a pleasure to record my debt to Professor T. A. Bennet-Clark, F.R.S., for his critical comments on every chapter, to Professor Sir Hans Krebs, F.R.S., for discussions on thermodynamics and to them both for at different times acting as teacher and mentor.

Thanks are extended to Mrs Howarth for typing the manuscript and Miss Janet Bloom for preparing the figures and finally to my wife for her constant encouragement.

PREFACE

BIOCHEMISTRY can be discussed in terms of a number of discrete topics such as carbohydrate, protein and lipid metabolism. However, with the passing of every year, it becomes increasingly difficult to define the boundaries of these topics, because some intermediates of carbohydrate metabolism are also intermediates in the metabolism of proteins, fats and nucleic acids. The term intermediary metabolism avoids the restrictions of a rigid classification of topics and denotes the more dynamic aspects of biochemistry.

The experimental study of intermediary metabolism is based on two principles. The first was stated in 1913 by Hopkins: 'the fact that the body, though the seat of a myriad reactions and capable perhaps of learning to a limited extent and under stress of circumstances, is in general able to deal only with what is customary to it.' This principle forms the theoretical basis for feeding experiments which have greatly increased in number since ^{14}C-labelled compounds have been available. It is interesting to note that in the eighteenth century it was believed that compounds synthesised by living organisms could not be synthesised in the laboratory. In the nineteenth century many natural products were synthesised and in the twentieth century chemists prepared labelled compounds which they, but not the plant, could distinguish from the normal.

The second principle is that, in general, the reactions observed in isolated enzyme systems reflect physiological events and are not artefacts of isolation. Occasionally terrorists suggest that enzymes are artefacts and less militant terrorists argue that the activity demonstrated *in vitro* may not reflect the *in vivo* function. However, results obtained with cell-free extracts have generally been concordant with results obtained by isotopic, spectrophotometric and genetic studies on intact cells and tissues. Nevertheless, this general principle must be applied with discretion, for as D. E. Green has pointed out, a mechanic could separate the parts of a car and use them to make a vacuum cleaner and he might even be tempted into believing

vii

that the parts he used were designed for the construction of a vacuum cleaner.

Progress in the study of intermediary metabolism is extremely rapid and the presentation of factual knowledge requires encyclopaedic treatment. This book represents an attempt to outline the concepts and theories of intermediary metabolism, using specific examples only to avoid becoming lost in generalisations. Topics chosen for discussion represent an arbitrary selection, the two major omissions being photosynthesis and nucleic acid metabolism. Fortunately photosynthesis has been the recent subject of three excellent monographs (Hill and Whittingham, 1955; Bassham and Calvin, 1957; Aronoff, 1957). There has, as yet, been little work on the metabolism of nucleotides and nucleic acid in plants, but excellent reviews of work with animals and bacteria have recently appeared (Buchanan and Standish, 1959; Hartman and Buchanan, 1959).

A few of the views expressed in this book are novel and many are controversial. It is hoped that the student will not accept any of these views unless they are in accordance with his own experience and thought. The book is based on a course of lectures given to the honours botany students of London University, but it is hoped that it will also provide students in other subjects with an introduction to the dynamic aspects of biochemistry.

D. D. DAVIES

KING'S COLLEGE
LONDON
November 1959

ABBREVIATIONS AND FORMULAE

The complexity of biochemical compounds necessitates the use of abbreviations. Trivial names such as diphosphopyridine nucleotide, cozymase and coenzyme I, as well as abbreviations such as DPN and CoI, are used to describe the same compound.

Abbreviations used in the text are listed below.

AMP, GMP, UMP, CMP, ITP	The 5′ phosphates of ribosyladenine, guanine, uridine, cytidine and hypoxanthine
AD , GDP, etc.	The 5′ (pyro) diphosphates of adenosine, guanine, etc.
ATP, GTP, etc.	The 5′ (pyro) triphosphates of adenosine, guanosine, etc.
CoA or CoASH	Coenzyme A
DPN (or DPN⁺)	Diphosphopyridine nucleotide
DPNH	Reduced form of DPN
TPN (or TPN⁺)	Triphosphopyridine nucleotide
TPNH	Reduced TPN
FAD	Flavin adenine dinucleotide
THFA	Tetrahydrofolic acid
Pi	Inorganic *ortho*phosphate

Formulae of some of the more important cofactors are displayed on the following pages.

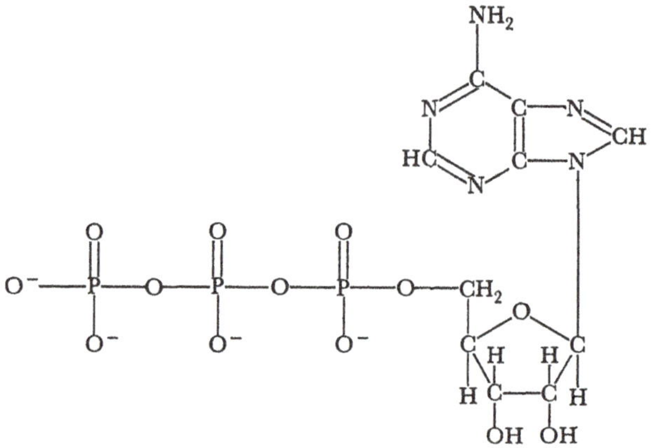

[1] Adenosine-5′ -triphosphate

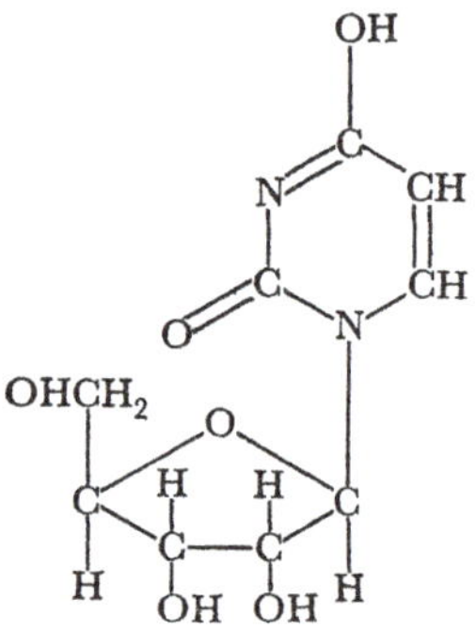

[2] Uridine

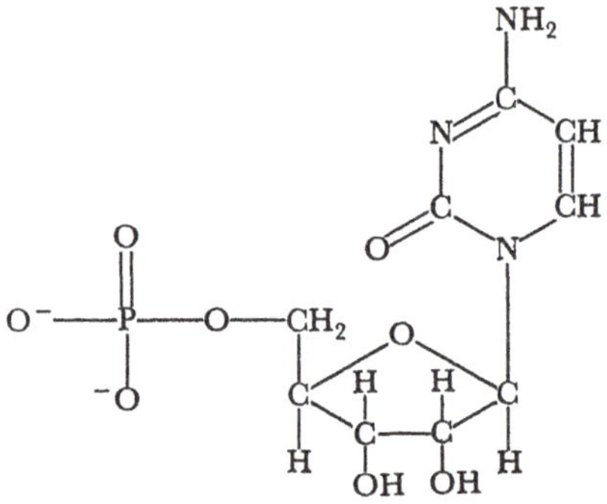

[3] Cytidine-5′ -monophosphate

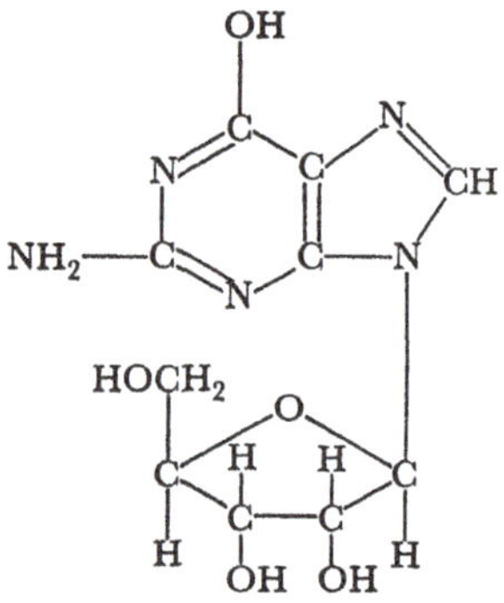

[4] Guanosine

x

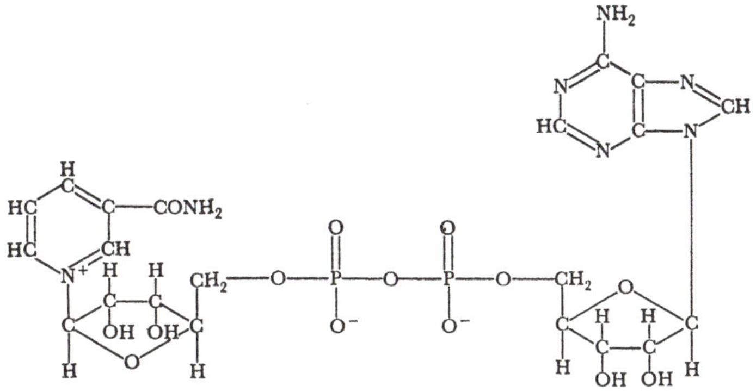

NH$_2$

C

N

C

N

CH

HC

C

N

H

C

C

CONH$_2$

HC

HC

CH

N$^+$

H

CH$_2$

O

P

O

P

O

CH$_2$

O

O

O

O$^-$

O$^-$

C

H

H

C

OH OH

C

H

H

O

C

H

C

C

OH OH

H

[5] Diphosphopyridine nucleotide

NH$_2$

C

N

C

N

CH

HC

C

N

H

C

C

CONH$_2$

HC

HC

CH

N$^+$

H

CH$_2$

O

P

O

P

O

CH$_2$

O

O

O

O$^-$

O$^-$

C

H

H

C

OH OH

C

H

H

O

C

H

H

C

C

OH

O

P

O$^-$

O$^-$

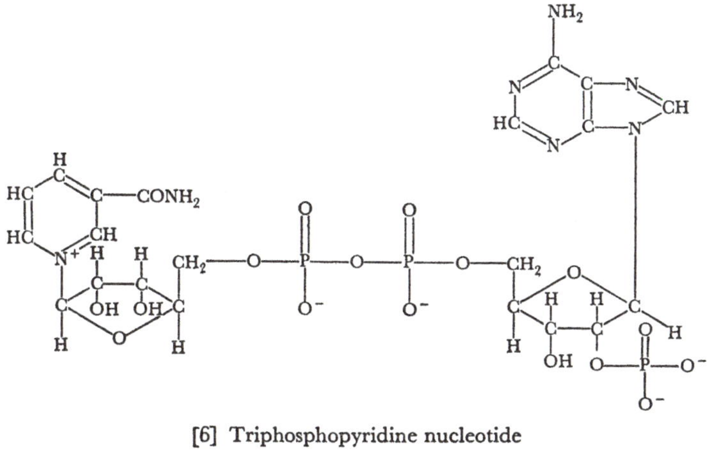

[6] Triphosphopyridine nucleotide

[7] Flavin adenine dinucleotide

[8] Coenzyme A

CHAPTER I

Metabolic Patterns and Cellular Organisation

METABOLISM involves a large number of chemical events, catalysed by enzymes and organised into metabolic patterns by multi-enzyme systems. The organisation of multi-enzyme systems may involve structural organisation—a viewpoint developed by Peters (1929, 1952) and more recently by Green (1957). However, as first pointed out by Hopkins (1932) and further developed by Dixon (1949), a high degree of organisation is a consequence of the chemical specificity of enzymes. Organisation in a factory is achieved by placing machines in a fixed sequence along a conveyor belt and it is the spatial relationship of the machines which produces the ordered sequence of operations. In contrast, enzymes may be in free solution and show what Dixon (1949) has called 'organisation by specificity'. The substrate specificity of enzymes ensures that the product of one enzyme is a substrate for only one other enzyme and so directs the substrate from enzyme to enzyme forming a linear metabolic pattern which branches when two or more enzymes can act on the same substrate. Available evidence indicates that with the exception of hexokinase (Saltman, 1953) all the glycolytic enzymes are soluble and the organisation of the multi-enzyme system of glycolysis can be attributed to the chemical specificity of the enzymes involved. Green (1957), on the other hand, maintains that it is difficult to reconcile the efficiency of glycolysis with an apparent lack of structural organisation, and postulates a close physical association between the glycolytic enzymes.

The understanding of the organisation of a multi-enzyme system requires a knowledge of the kinetics of the system. The kinetic analysis of a metabolic chain such as glycolysis is extremely complex, but is somewhat simplified when a steady state is maintained. The term 'steady state' used in biochemistry is equivalent to the term 'stationary state' employed by chemists and refers to the state in which the rate of utilisation of substrate is constant and equal to the rate of formation of product and in which the concentration of intermediates remains constant. If we consider a simple reaction such as the generalised mechanism of enzyme reaction proposed by Briggs and Haldane (1925)

$$E + S \rightleftharpoons ES \rightleftharpoons E + P$$

we can distinguish a number of stages as shown in fig. 1.

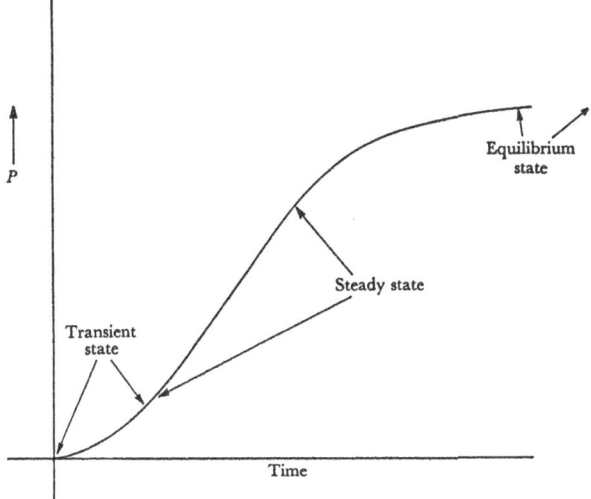

Fig. 1. Course of an enzyme-catalysed reaction.

Initially we may distinguish the transient state in which there is a rapid disappearance of free enzyme (E) as it combines with the substrate to form ES. The detection of the transient state requires special apparatus because of the very rapid rate of reaction between enzyme and substrate. The transient state leads into the steady state in which the rate of formation of P and the concentration of the intermediate ES remain constant. The rate of formation of P eventually declines and the reaction reaches the equilibrium state. This tendency for a steady state to pass into an equilibrium state is a property of *closed* systems. A system is termed 'closed' when no material exchange takes place between the system and the environment. Thus experiments with enzymes performed in test tubes, Warburg flasks or spectrophotometer cells show the kinetics of closed systems. The kinetics of metabolic processes taking place in the living plant

2

are those of an 'open' system. When a material exchange occurs between a system and its environment, the system is termed 'open'. For example, if sugar diffuses from the vacuole into the cytoplasm of the cell and is there oxidised to carbon dioxide, the kinetics of an open system are applicable. Some important features of open systems can be presented by consideration of the simple system shown below.

$$A_o \xrightarrow{\; k_1 \;} \bigg| \to A_i \underset{k_3}{\overset{k_2}{\rightleftharpoons}} B_i \xrightarrow{\; k_4 \;} \bigg| \to B_o$$

$$\bigg| \leftarrow \text{membranes} \to \bigg|$$

The system includes two diffusion constants k_1 and k_4. In the special case where k_1 and $k_4 = 0$, the system is closed and A_i and B_i tend to approach chemical equilibrium. If k_1 and k_4 are large, material passes rapidly through the system whilst A_i and B_i approach constant concentrations characteristic of the steady state. An important feature of an open system is that it is self-regulatory, in the sense that it follows the principle of Le Chatelier and when a constraint is applied the system reacts to produce a new steady state. Consider the case in which k_4 is increased; the internal concentration B_i tends to fall, but this leads to an increased production of B from A due to the slower back reaction $(k_3 B_i)$. Similarly, the increased rate of formation of B from A tends to lower the concentration of A which in turn is offset by the increased rate at which A enters the system and thus establishes a new steady state.

During the transition from one steady state to another the concentration of an intermediate may change exponentially or may fall below the original steady-state concentration before rising to the new level ('false start') or may rise above and subsequently fall to the new steady-state level ('overshoot'). Fluctuations have been observed during studies of photosynthesis as, for example, when the partial pressure of carbon dioxide is changed (fig. 2). Much of the interest in the kinetics of metabolic reactions has been directed towards the concept of a rate-determining step, originally developed by Blackman (1905) and expressed in the oft-quoted statement that 'when a process is conditioned as to its rapidity by a number of separate factors, the rate of the process is limited by the pace of the slowest factor'. This concept has had a most influential role on subsequent thought, but requires modification to be acceptable in terms of present-day kinetic theory. An extension of Blackman's views by Crozier (1924) to interpret the temperature relationships of biological activity in terms of a 'master reaction' which could be detected by means of its Arrhenius constant, has been severely criticised (see Hearon, 1952). On the other hand,

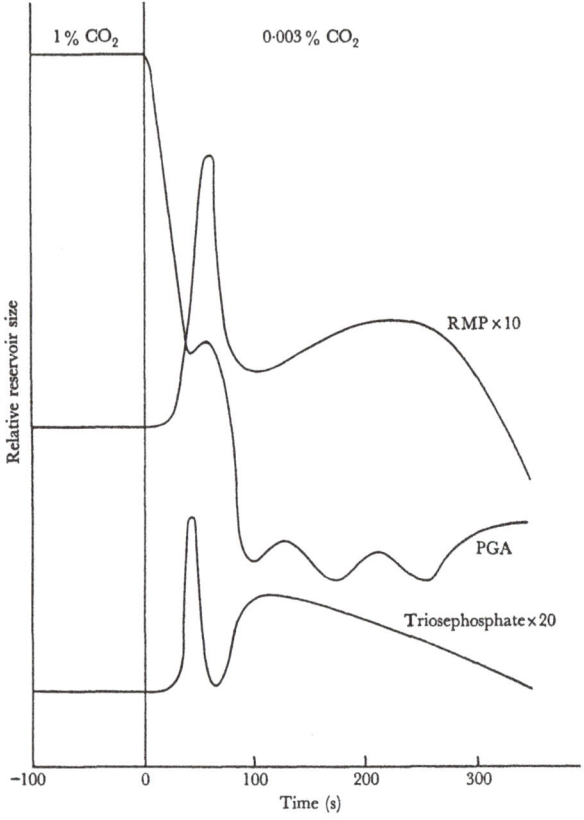

Fig. 2. Effect of changes in the partial pressure of carbon dioxide on the concentration of some photosynthetic intermediates. Temperature 6°. Modified from Wilson and Calvin (1955). RMP is ribose-5-phosphate, PGA is phosphoglyceric acid.

many textbooks of botany and plant physiology present Blackman's analysis of the law of limiting factors without the modifications which subsequent kinetic analysis has shown to be necessary. The continued acceptance of these views is perhaps due to the acceptance of a false analogy. For example, if a wall is being erected by a bricklayer and his labourer, the rate of building will be determined by the pace of the slower worker.

4

This example agrees with Blackman's views because the pace of the slower worker is not affected by the pace of the faster; thus if the bricklayer is a slow worker, bricks may accumulate around him but this will not increase his rate of bricklaying. Chemical reactions, on the other hand, obey the law of Mass Action, the rate being proportional to the concentration (or more correctly the activity) of the substrate. Consequently the overall rate of two consecutive chemical reactions is determined by the velocity constants of both reactions. Consider consecutive first-order reactions

$$A \xrightarrow{k_1} B \xrightarrow{k_2} C.$$

If A_o is the concentration of A at time t_o and C_A, C_B and C_C are the values for A, B and C respectively at time t, then

$$A_o = C_A + C_B + C_C$$

and
$$-\frac{dC_A}{dt} = k_1 C_A, \quad \frac{dC_B}{dt} = k_1 C_A - k_2 C_B$$

$$\frac{dC_C}{dt} = k_2 C_B,$$

from which it can be shown that

$$C_C = A_o \left[1 - \frac{1}{k_1 - k_2} (k_1 e^{-k_2 t} - k_2 e^{-k_1 t}) \right].$$

It is clear that the rate of formation of C depends upon k_1 and k_2.

If we consider a series of reactions in the steady state, all the reactions have the same velocity, so that no reaction can be regarded as the slowest. However, it is possible for certain reactions to be rate-determining and so they may be regarded as 'pacemakers'. Consider the reaction sequence

$$A_o \longrightarrow| \rightarrow A \rightleftharpoons B \rightleftharpoons C \rightleftharpoons D \longrightarrow| \rightarrow D_o.$$

The overall rate is a function of the velocity constants of all the steps and the rate increases to a maximum as the concentration of material supplied (A_o) is increased. The maximum rate is attained when one of the enzymes shows zero-order kinetics, that is, when one of the enzymes is saturated with its substrate. The relationship between the rate of an enzyme reaction and the concentration of substrate is shown in fig. 3. The curve shown in fig. 3a is asymptotic, the maximum velocity being attained at

5

infinite substrate concentration (Fig. 3 *b*). However, at high substrate concentration the rate is virtually independent of substrate concentration and the reaction can be considered zero order. In a series of reactions, the first reaction showing zero-order kinetics acts as the pacemaker for the whole process.

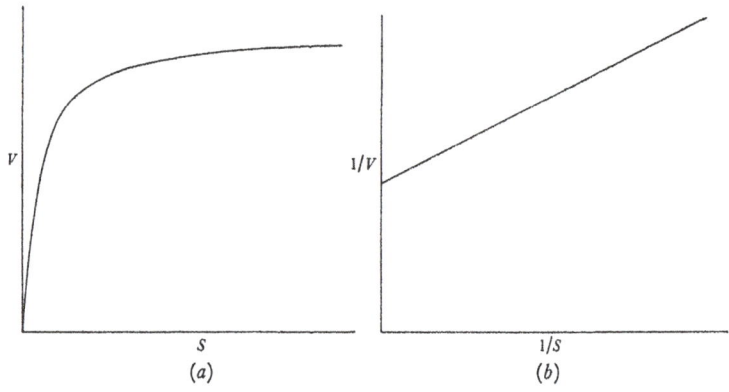

Fig. 3. Effect of substrate concentration on the velocity of an enzyme-catalysed reaction. The relationship between velocity *v* and substrate concentration *S* is found in many cases to conform to the Michaelis equation

$$v = \frac{V_{max}S}{K_m + S}; \quad \text{when} \quad \frac{v}{V_{max}} = \tfrac{1}{2}, \quad K_m = S.$$

(V_{max} represents the velocity at infinite substrate concentration.)

Rearranging the Michaelis equation gives

$$\frac{1}{v} = \frac{K_m}{V_{max}}\frac{1}{S} + \frac{1}{V_{max}}.$$

Thus, plotting $1/v$ against $1/S$ gives a straight line which cuts the vertical axis at $1/V_{max}$ and has a slope K_m/V_{max}.

The rate of an irreversible reaction is unaffected by the concentration of its products; thus reactions subsequent to an irreversible reaction cannot affect the rate, but the first irreversible reaction of a sequence may be rate-determining. The rate of an irreversible reaction is of course affected by previous reactions; thus if a zero-order reaction precedes an irreversible reaction the overall rate is determined by the zero-order reaction. Should an irreversible reaction precede a zero-order reaction, a steady state is impossible.

These considerations are valid when applied to a linear series of reactions, but as subsequently shown the limitations imposed

6

by irreversibility do not apply in more complex metabolic patterns.

Applying these principles to the glycolytic sequence (fig. 4), it might at first appear that hexokinase could be rate-determining by reason of its irreversibility[1] ($\Delta G = -5.7$ kcal).

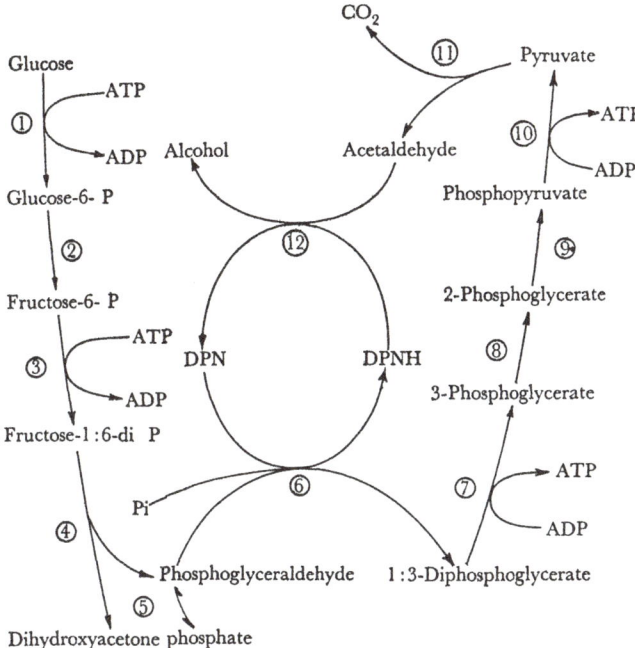

Fig. 4. Reactions of the Embden–Meyerhof–Parnas pathway (glycolytic pathway). The enzymes are numbered as follows: 1, hexokinase; 2, hexose phosphate isomerase; 3, phosphofructokinase; 4, aldolase; 5, triosephosphate isomerase; 6, triosephosphate dehydrogenase; 7, 3-phosphoglycerate kinase; 8, phosphoglyceromutase; 9, enolase; 10, pyruvate kinase; 11, pyruvic decarboxylase; 12, alcohol dehydrogenase.

Low concentrations of glucose-6-phosphate inhibit hexokinase isolated from brain and heart, hence in these tissues the rate at which glucose is phosphorylated depends on the rate at which glucose-6-phosphate is removed. Hexokinase could, nevertheless, be rate-determining by reason of its low concentration or low activity, and evidence that the rate of glycolysis

[1] In theory all enzyme reactions are reversible, but when the free energy change is greater than 4–5 kcal, the back reaction is so small that it may be neglected.

7

by leucocytes is limited by hexokinase has been described by Beck (1958). Product inhibition of plant hexokinase has not been reported, but other considerations establish that the hexo-kinase reaction cannot be rate-determining by reason of its irreversibility. In the fermentation of glucose, two moles of ATP are used in the formation of fructose diphosphate and four moles of ATP are produced in subsequent reactions, giving a net gain of two moles of ATP. The rate at which glucose is phos-phorylated is thus dependent upon the rate at which subsequent reactions can generate ATP. The hexokinase reaction is thus

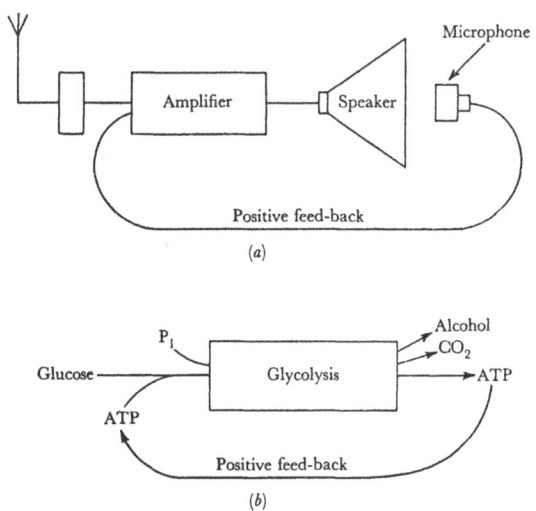

Fig. 5. Examples of positive feed-back mechanisms. (a) Electronic feed-back; (b) metabolic feed-back.

coupled to reactions producing ATP by what Krebs (1947) has called an 'enzyme cycle' and which in this particular case shows some of the properties of a positive feed-back mechanism. The general property of positive feed-back is that it accelerates a system to function at maximum capacity. A radio receiver picking up a weak signal has a low output, but if a microphone is placed in front of the loud speaker and connected to the gramophone input, part of the speaker output is fed back into the amplifier. The increased input produces a louder output and the positive feed-back ensures that the speaker output rapidly reaches a maximum which may be determined by any of the valves in the amplifier. Similarly, part of the output of glyco-

8

lysis (ATP) is fed back into the glycolytic system by the hexo-kinase reaction and so leads to the production of more ATP (fig. 5).

In general, enzyme cycles are not feed-back mechanisms, but rather catalytic cycles. For example, green leaves contain the enzymes glyoxylic reductase and glycolic acid oxidase required for the hydrogen-transporting system shown in fig. 6. The

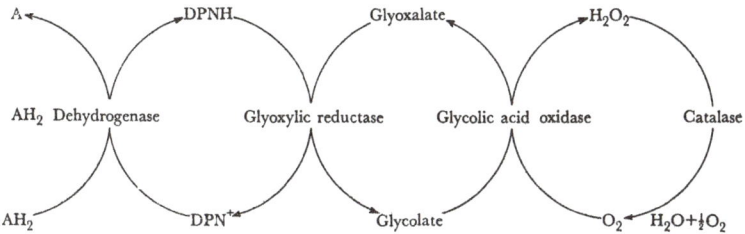

Fig. 6. Hydrogen-transporting system formed by linked enzyme cycles. Glyoxylic reductase may also be called glycolic dehydrogenase. Glycolic acid oxidase is a flavoprotein present in green leaves which reacts directly with oxygen. The quantitative importance of these reactions is uncertain and may be determined by the affinity of the flavoprotein for oxygen.

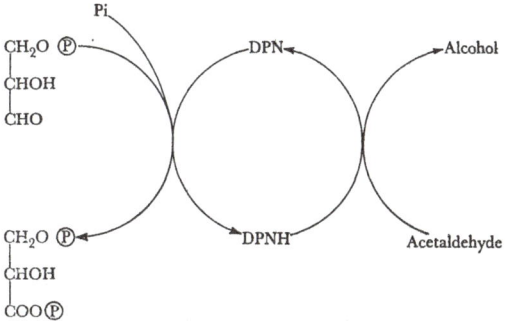

Fig. 7. The oxidation-reduction couple of glycolysis. By means of this enzyme cycle, the oxidative step of glycolysis is exactly balanced by the reduction of acetaldehyde.

operation of this pathway in cell-free homogenates has been demonstrated by Zelitch and Ochoa (1953), but the extent to which it participates *in vivo* is unknown. Perhaps the best known enzyme cycle is the oxidation-reduction couple of glycolysis (fig. 7).

Here the two enzymes are linked by the redox couple DPN/DPNH. It has often been implied that DPN and TPN act as mobile carriers between the various dehydrogenase

9

systems. Thus Parnas (1943) called the pyridine nucleotide coenzymes 'mobile enzymes' in contrast to firmly bound 'fixed enzymes' such as the flavin coenzymes. On the other hand, in biological systems where the number of binding sites is of the order of magnitude of available DPN molecules, interaction of apoenzymes with bound DPN could play an important role in physiological oxidation-reduction. Crystalline triosephosphate dehydrogenase contains firmly bound DPN which is not removed by dialysis. Nevertheless, the coenzyme appears to be available to other dehydrogenases. Cori, Velick and Cori (1950) observed that when the *bound* DPN of triosephosphate dehydrogenase was reduced it could be reoxidised by pyruvate and lactic dehydrogenase. Nygaard and Rutter (1956) have confirmed these findings and demonstrated that acetaldehyde and alcohol dehydrogenase react more readily with DPNH bound to triosephosphate dehydrogenase than with free DPNH. These observations suggest that two protein molecules can bind and activate a single coenzyme molecule. Such a ternary complex would produce serious problems of steric hindrance unless the enzymes can act on opposite sides of the nicotinamide ring.

The mechanism and stereospecificity of the pyridine nucleotide dehydrogenases has been studied by Vennesland and her collaborators (see Vennesland (1955)). Prior to the work of Vennesland it was widely believed that the mechanism of action of a dehydrogenase was essentially the transfer of electrons, the loss or gain of hydrogen atoms being explained by the release, or acceptance of protons from the solution. Vennesland established that there is a direct transfer of hydrogen from substrate to coenzyme. The alcohol dehydrogenase catalysed reduction of ethanol by DPN was carried out in heavy water (D_2O).

$$CH_3CH_2OH + DPN^+ \xrightleftharpoons{\;D_2O\;} DPNH + H^+ + CH_3CHO,$$ and the reduced DPN formed, was found to contain no deuterium. It follows that hydrogen, removed from the ethanol, did not equilibrate with the heavy water. This conclusion was confirmed by showing that the reduction of deuterium-labelled ethanol gave rise to reduced DPN with one deuterium atom per mole:

$$CH_3CD_2OH + DPN \xrightleftharpoons{\;[H_2O]\;} DPND + H^+ + CH_3CDO.$$

Vennesland established that the enzymic reduction of DPN and the enzymic oxidation of DPNH were stereospecific. Thus a particular dehydrogenase can add or remove hydrogen from only one side of the nicotinamide ring. When DPN is chemically reduced in heavy water, the two possible stereoisomers are formed:

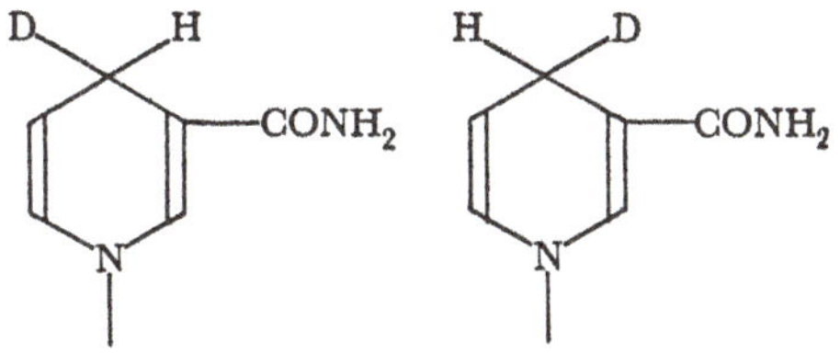

D, H   H, D

[ring]—CONH$_2$ [ring]—CONH$_2$

The stereospecific enzymic reduction of DPN by a deuterium-labelled substrate would produce only one of the two isomers of DPND. Assuming the same stereospecificity, the alcohol de-hydrogenase-catalysed oxidation of DPND would be expected to remove all the deuterium from the enzymically prepared DPN and half the deuterium from the chemically prepared DPND. Experimental findings confirmed the assumption of stereospecificity.

It was further demonstrated by Vennesland that the transfer of hydrogen to and from a substrate is stereospecific. DPND prepared enzymically was used in the following reaction:

$$CH_3CHO + H^+ + DPND \rightleftharpoons CH_3CHDOH + DPN^+.$$

If the reduction of acetaldehyde is stereospecific, the deuterium-labelled ethanol should be optically active. The cost of DPND precluded the preparation of sufficient deuterium-labelled ethanol to permit the use of a polarimeter to measure optical activity. Proof that only one isomer of deuterium-labelled ethanol had been formed was obtained by carrying out the reaction:

$$CH_3CHDOH + DPN^+ \rightleftharpoons CH_3CHO + DPND + H^+.$$

If the reaction is stereospecific, the acetaldehyde should contain no deuterium and such was the experimental finding. These results are summarised below:

(1) There is a *direct* transfer of hydrogen from the substrate to the coenzyme.

(2) The removal of hydrogen from the substrate is stereospecific. Thus, for example, the enzyme alcohol dehydrogenase discriminates between the two methylene hydrogens of alcohol.

(3) The transfer of hydrogen to position 4 of the nicotinamide ring of DPN is stereospecific. Alcohol dehydrogenase transfers hydrogen to one side of the ring which is nominally called the 'A' side and is used as a reference to compare the stereospecificity of other dehydrogenases (see table 1).

TABLE I. *Steric specificity for* DPN (after Levy and Vennesland, 1957)

Dehydrogenase	Source	Specificity
Alcohol (ethanol)	Yeast, wheat germ	A
Acetaldehyde	Liver	A
L-Lactate	Heart	A
L-Malate	Heart, wheat germ	A
D-Glycerate	Spinach	A
α-Glycerophosphate	Yeast, muscle	B
L-Glutamate	Liver	B
3-Phosphoglyceraldehyde	Yeast, muscle	B
DPNH cytochrome c	Liver, heart	B

It will be observed that triosephosphate dehydrogenase transfers hydrogen to the B side of the ring, whereas alcohol and lactic dehydrogenases transfer to the opposite side. It is tempting to speculate that the two dehydrogenases of glycolysis form a complex which permits the hydrogen from phosphoglyceraldehyde to 'flow over' the shared coenzyme to acetaldehyde. The view that certain dehydrogenases are intimately connected *in vivo* is open to experimentation. Thus if lactic and triosephosphate dehydrogenase are associated so that hydrogen from lactate flows over the coenzyme to enter phosphoglyceraldehyde, then feeding 2-deuteriolactate should label the aldehyde group of phosphoglyceraldehyde. In consequence, deuterium would be expected to enter reserve carbohydrates, formed by a reversal of the glycolytic sequence. If there is no association between dehydrogenases, specific labelling of carbohydrates would not be expected. Experiments along these lines have not been carried out with plants, but animal-feeding experiments have been reported by Hoberman (1958). When 2-deuteriolactate was fed to rats, liver glycogen was found to be labelled mainly in C-4 of the glucose residues, with smaller amounts of deuterium in C-6. The observed labelling of C-4 is in accord with the view that because of enzyme association deuterium enters the aldehyde of phosphoglyceraldehyde, but fails to label C_3 of glucose because deuterium entering dihydroxyacetone phosphate would equilibrate with water. The labelling of C-6 suggests transfer of label to malate and equilibration with fumarate to give deuteromalate. It would be of interest to determine the steric specificity of malic enzyme which should be opposite to the specificity of lactic dehydrogenase, functioning with TPN, if the above views are valid.

In contrast to these findings, we note that Velick (1958) could not obtain evidence for a complex between triosephosphate dehydrogenase, lactic dehydrogenase and reduced DPNH, and emission spectra and polarisation values are in accord with the view that there is a transfer of coenzyme from one protein to another.

A metabolic cycle occurs when one of the products of a bi-molecular reaction is converted, by a number of steps, into one of the substrates of the bimolecular reaction. The Krebs tricarboxylic acid cycle (fig. 8) is the most fully investigated metabolic cycle known. In one turn of the cycle acetyl-CoA and oxaloacetate combine to give citrate and CoA, the citrate undergoing a number of reactions involving the production of two moles of carbon dioxide and the regeneration of oxalo-acetate. The cycle functions as a catalytic respiratory mechanism and cannot produce a net increase in organic acid. Inter-mediates of the cycle are, however, important in many synthetic reactions, for example, α-oxoglutarate is a precursor of gluta-mate and related amino acids whilst succinate is a precursor of chlorophyll. The effect of removing cycle intermediates would be to slow down and eventually stop the cycle. Essentially the same view has been put forward by Vishniac, Horecker and Ochoa (1957) who argue that 'the assumption that di- and tri-carboxylic acids are formed by reactions of the Krebs cycle, or are derived from Krebs cycle acids, is incompatible with the simultaneous functioning of such a cycle as a catalytic respira-tory mechanism'. Krebs (1941), on the other hand, has pointed out that acids may be removed from the cycle without impairing its catalytic function, provided there is an equivalent fixation of carbon dioxide into oxaloacetate and malate.

The effect of cycling is to convert a product into a substrate and consequently an irreversible reaction cannot determine the rate of cycling, but does determine the direction of cycling. A zero-order reaction can, however, determine the rate of cycling. In the absence of a zero-order reaction, the rate of cycling will increase with increasing concentration of acetyl-CoA,

13

but the increase will be limited by the total concentration of cycle acids. The limitation is due to the fact that an increase in citrate concentration necessitates a fall in the concentration of oxaloacetate. The observation (Bennet-Clark and Bexon (1943)) that the rate of respiration of beet disks is greatly increased by an external supply of organic acids, suggests that

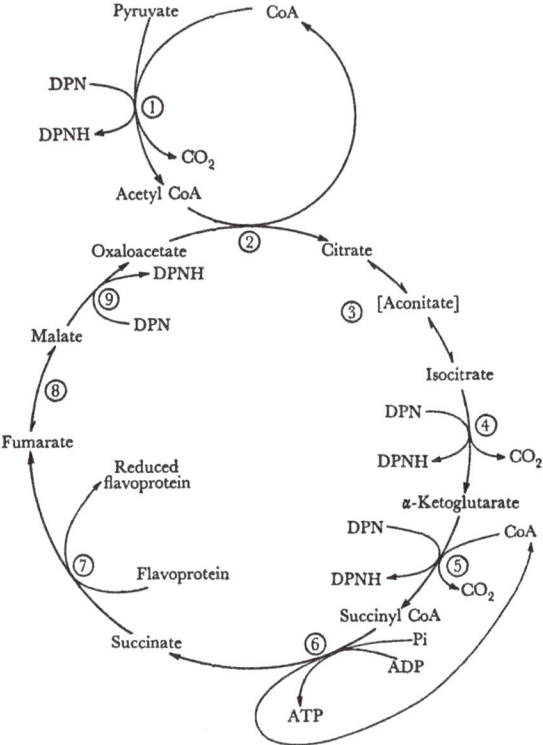

Fig. 8. Reactions of the tricarboxylic acid cycle (Krebs cycle). The enzymes are numbered as follows: 1, pyruvic dehydrogenase; 2, condensing enzyme; 3, aconitase; 4, DPN-*iso*citric dehydrogenase; 5, α-ketoglutaric dehydrogenase; 6, 'P' enzyme; 7, succinic dehydrogenase; 8, fumarase; 9, malic dehydrogenase.

under certain conditions the rate of cycling may be limited by the concentration of acids at the site of metabolic activity. A reduction in the concentration of acetyl-CoA would be expected to lead to an accumulation of oxaloacetate. However, malic dehydrogenase shows product inhibition (Davies and Kun, 1957), so that at concentrations above 10^{-4}M oxaloacetate inhibits its own synthesis. The fact that oxaloacetate also inhibits succinic dehydrogenase (Pardee and Potter, 1948; Avron and Biale, 1957) suggests the possibility that the concentration of oxaloacetate may regulate the rate of cycling in much the same way as a thermostat controls temperature (fig. 9). If the temperature rises above a preset value, the toluene expands, raises the mercury level and so completes the circuit to the relay which opens the heater circuit. The heater then remains 'off' until the temperature falls and reverses the process.

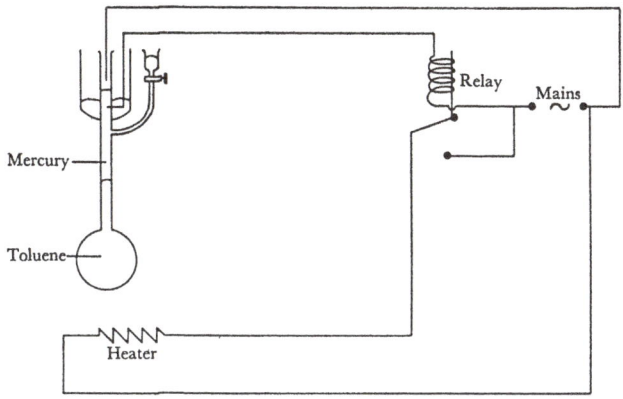

Fig. 9. Control of temperature as a negative feed-back mechanism.

This type of control is an example of negative feed-back (see Krebs, 1957) which may be defined in terms of metabolism as an arrangement whereby a product of a reaction sequence inhibits one of the earlier reactions in the sequence, thereby slowing down the overall rate until the product is removed. It seems probable that further investigations will establish negative feed-back as a widespread control mechanism. An interesting example is the control of *iso*leucine synthesis in *Escherichia coli*. Starting from aspartate the biosynthesis of *iso*leucine may be represented as in fig. 10.

Threonine deaminase is inhibited by *iso*leucine and because the enzyme has a much greater affinity for the inhibitor than for its substrate (Umbarger and Brown, 1958) a small increase in the concentration of *iso*leucine will effectively block the deamination of threonine. Threonine inhibits homoserine kinase (Wormser and Pardee, 1958) so that two feed-back mechanisms appear to be co-ordinated to control the biosynthesis of *iso-*

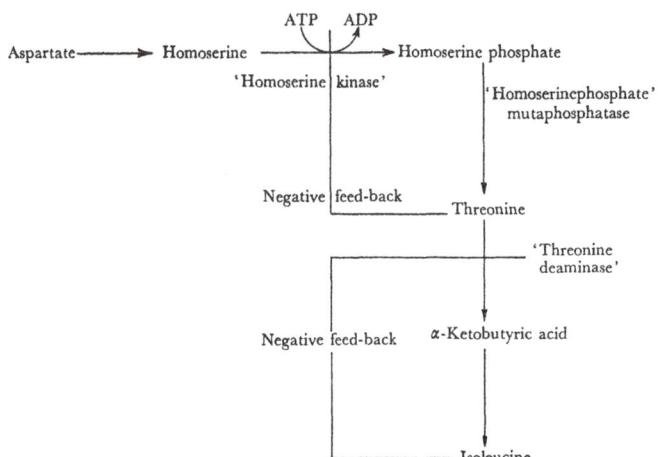

Fig. 10. Control of *iso*leucine biosynthesis by negative feed-back.

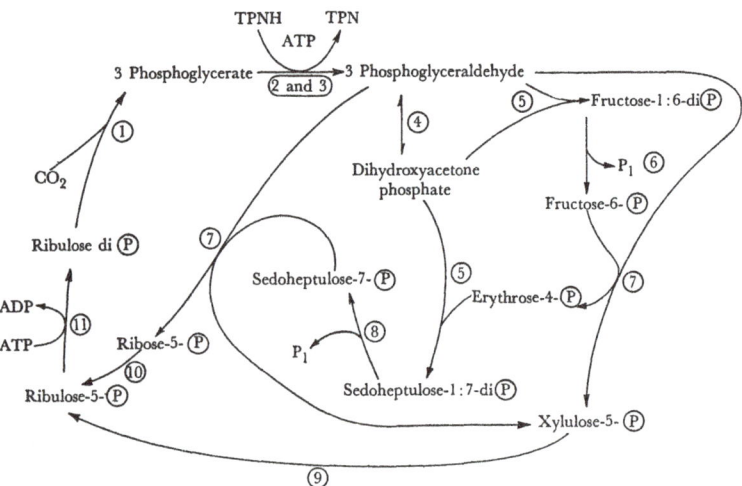

Fig. 11. Reactions of the photosynthetic cycle (reductive pentose-phosphate cycle). The enzymes are numbered as follows: 1, carboxydismutase; 2, phosphoglycerate kinase; 3, triosephosphate dehydrogenase; 4, triosephosphate isomerase; 5, aldolase; 6, fructose-1:6-diphosphatase; 7, transketolase; 8, sedoheptulose-1:7-diphosphatase; 9, phosphoketopentose epimerase; 10, phosphoribose isomerase; 11, ribulose phosphate kinase.

leucine. It should be noted that negative feed-back by end-product inhibition is most effective when the reaction inhibited is at the beginning of a metabolic sequence.

Unlike the Krebs cycle, the photosynthetic cycle (fig. 11) can

16

produce a net increase in cycle intermediates, because the product of the photosynthetic cycle (3-phosphoglycerate) is also a cycle intermediate. This circumstance produces a clear example of positive feed-back which functions to accelerate the cycle to a maximum rate. In darkness, the steady-state concentration of the photosynthetic cycle intermediates is low (fig. 12 and

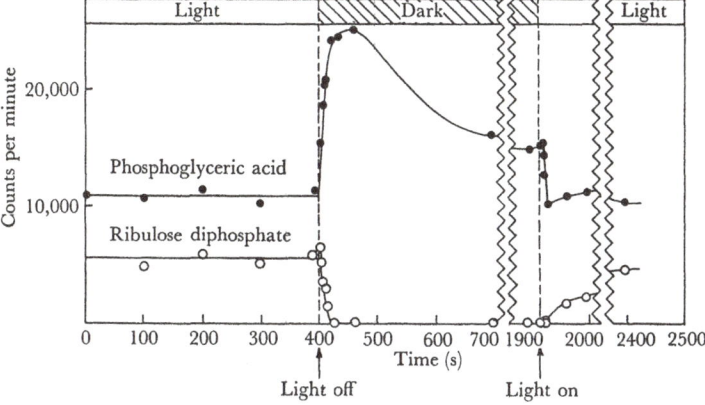

Fig. 12. Effect of light and dark on the concentration of phosphoglycerate and ribulose dephosphate.

TABLE 2. *Steady-state concentration of phosphosynthetic intermediates and related hexose phosphates (in μmoles/ml of algae)*

Compound	Dark	Light	Change
Phosphoglyceric acid	2·42	1·63	−0·79
Ribulose diphosphate	< 0·005	0·51	+0·51
Pentose monophosphates	0·08	0·17	+0·09
Sedoheptulose diphosphate	0·006	0·006	0
Fructosediphosphate	0·001	0·004	+0·003
Triosephosphate (area)	0·12	0·21	+0·09
Fructose-6-phosphate	0·12	0·12	0

table 2), but when the light is switched on, carbon dioxide is fixed and the cycle produces phosphoglycerate which is fed back into the cycle, producing a net increase in the concentration of intermediates and accelerating the cycle to a maximum rate.

Two possible variants of the photosynthetic cycle are shown in figs. 11 and 13. The cycle shown in fig. 11 does not assign a role to the enzyme transaldolase, but does account for the observed formation of sedoheptulose-1:7-diphosphate and requires the presence of the enzyme sedoheptulose diphosphatase which is known to be present in chloroplasts (Racker and Schroeder, 1958). The cycle shown in fig. 13 does not account for the formation of sedoheptulose-1:7-diphosphate and requires the presence of transaldolase. It is not yet possible to decide which cycle represents the reactions occurring within the chloroplasts and both may participate. The cycle represented in fig. 11 explains many experimental observations and a choice

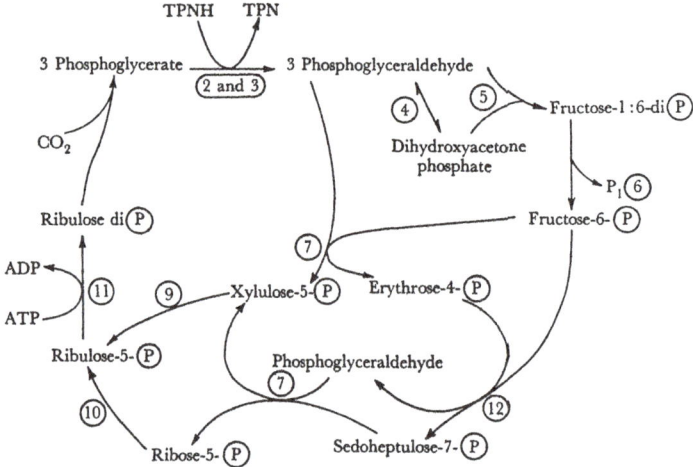

Fig. 13. Reactions of the photosynthetic cycle. The enzymes are numbered as in fig. 11 (12 is transaldolase).

in favour of this scheme could be made if the absence of transaldolase from chloroplasts could be established.

The photosynthetic cycle participates in the reduction of carbon dioxide to carbohydrates and has been called the reductive pentose-phosphate cycle in contrast to the oxidative pentose-phosphate pathway (fig. 14) which effects the complete oxidation of glucose-6-phosphate. The reactions shown in fig. 14 involve two oxidative steps, only one of which is a decarboxylation. That all six carbons of glucose can be released as carbon dioxide is due to the combined action of the non-oxidative enzymes which can transfer all the carbon atoms of glucose to the 1-position of 6-phosphogluconate. The enzyme

18

transketolase, which catalyses two of the reactions shown in fig. 14, can also catalyse the reaction

sedoheptulose-7-phosphate + erythrose-4-phosphate
$\rightleftharpoons$ fructose-6-phosphate + ribose-5-phosphate.

Consequently the reactions of the pentose-phosphate pathway can be arranged to produce the cycle shown in fig. 15. An interesting feature of this metabolic pattern is that it shows a cycle within a cycle, since half the ribose-5-phosphate which reacts with xylulose-5-phosphate is regenerated by subsequent reactions.

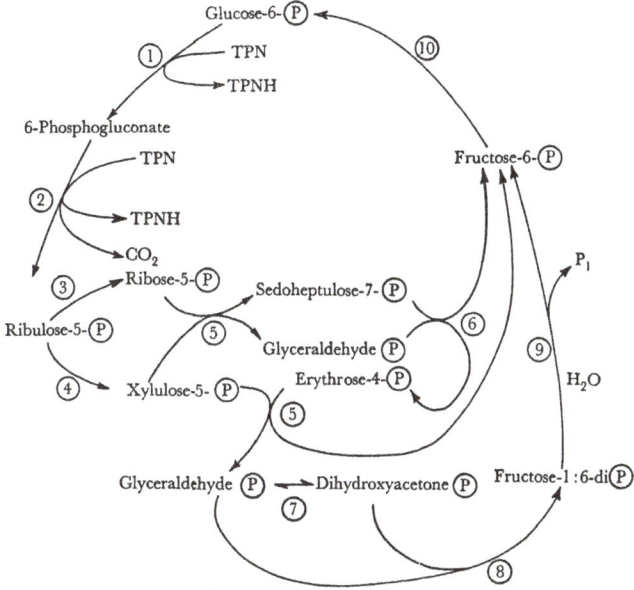

Fig. 14. Reactions of the reductive pentose-phosphate pathway. The enzymes are numbered as follows: 1, glucose-6-phosphate dehydrogenase; 2, 6-phosphogluconic dehydrogenase; 3, phosphoribose isomerase; 4, phosphodetopentose isomerase; 5, transketolase; 6, transaldolase; 7, triosephosphate isomerase; 8, aldolase; 9, fructose-1:6-diphosphatase; 10, hexosephosphate isomerase.

The term cycle is sometimes loosely applied as, for example, in referring to the β-oxidation of fatty acids as the fatty-acid cycle. Lynen (1954) has pointed out that the metabolic pattern of β-oxidation forms a helix rather than a spiral. The application of the term cycle to the reactions of the pentose-phosphate pathway may represent an over-formalisation of the reactions involved.

Available evidence suggests that the enzymes of the pentose-

2-2

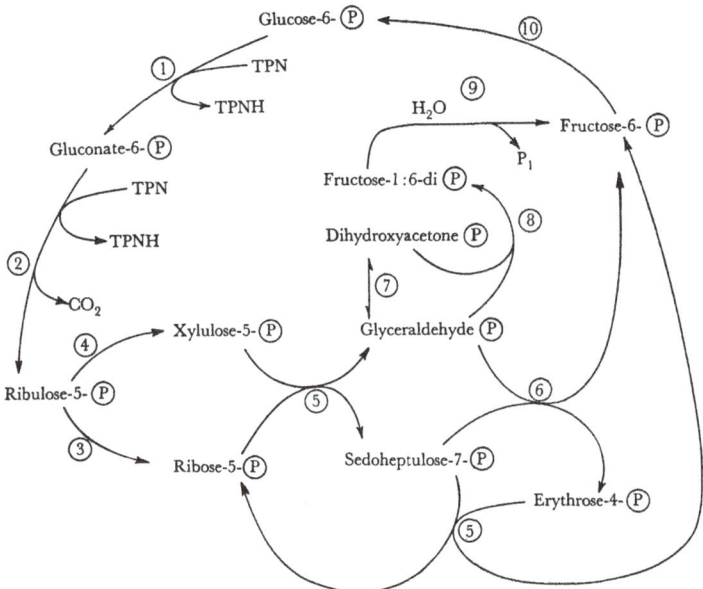

Fig. 15. Reactions of the pentose-phosphate cycle. The enzymes are numbered as in fig. 14.

TABLE 3. *Substrate specificity of some enzymes participating in the pentose-phosphate pathway*

A. TRANSKETOLASE

Glycolaldehyde donors	Glycolaldehyde acceptors
Hydroxypyruvate	Glycolaldehyde
L-Erythrulose	DL-Glyceraldehyde
Xylulose-5-phosphate	Phosphoglyceraldehyde
Fructose-6-phosphate	Erythrose-4-phosphate
Sedoheptulose-7-phosphate	Ribose-5-phosphate
Octulose-8-phosphate	Allose-6-phosphate

B. TRANSALDOLASE

Dihydroxyacetone donors	Dihydroxyacetone acceptors
Sedoheptulose-7-phosphate	Phosphoglyceraldehyde
Fructose-6-phosphate	Ribose-5-phosphate

C. ALDOLASE

Carbonyl compound	Compound with activated methylene group	Product
D-Erythrose	Dihydroxyacetone	Sedoheptulose-1-phosphate
D-Glyceraldehyde	phosphate	Fructose-1-phosphate
L-Glyceraldehyde	—	Sorbose-1-phosphate
	—	
D-Phosphoglyceraldehyde	—	Fructose-1:6-diphosphate
Glycolaldehyde	—	Xylulose-1-phosphate
Phosphoglycolaldehyde	—	Xylulose-1:5-diphosphate

20

phosphate pathway are in free solution. Organisation in free solution is possible by means of enzyme specificity, but some of the enzymes involved in the pentose-phosphate pathway exhibit a wide range of substrate specificity (see table 3). This lack of specificity makes it possible to draw on paper two distinct pentose-phosphate cycles, both of which can account for the complete oxidation of glucose-6-phosphate. It follows that if the enzymes of the pentose-phosphate cycle are in free solution they cannot produce an ordered sequence of events such as those represented in a formal cycle. The pentose-phosphate pathway may be thought of as a pool of all the substrates shown in fig. 14 maintained in a state of dynamic equilibrium. The equilibrium constants for all the reactions involved have not yet been determined, but from results published so far (table 4) it would appear that most of the reactions other than the two dehydrogenations are readily reversible. Consequently if any intermediate is removed from the pool, the concentration of all the reactants will readjust to a new equilibrium value.

TABLE 4. *Equilibrium constants for some of the reactions involved in the pentose-phosphate pathway*

Enzyme	Equilibrium	K
Transaldolase	$\dfrac{\text{(Fructose-6-phosphate) (tetrose-phosphate)}}{\text{(Sedoheptulose-7-phosphate) (glycerald-phosphate)}}$	0·8
Transketolase	$\dfrac{\text{(Fructose-6-phosphate) (glycerald-3-phosphate)}}{\text{(Xylulose-5-phosphate) (erythrose-4-phosphate)}}$	1
Ribose-5-phosphate isomerase	$\dfrac{\text{Ribose-5-phosphate}}{\text{Ribulose-5-phosphate}}$	3
Xylulose-5-phosphate	$\dfrac{\text{Xylulose-5-phosphate}}{\text{Ribulose-5-phosphate}}$	3
Phosphohexose	$\dfrac{\text{Glucose-6-phosphate}}{\text{Fructose-6-phosphate}}$	2·3

The metabolic pattern discussed on previous pages are facets of the metabolic whole. In some cases a metabolic pattern is intimately associated with a cellular structure which can be isolated, and so permit the analysis of an isolated metabolic pattern. Thus, for example, a detailed study of the Krebs cycle is possible because of its association with mitochondria. Within the cell the various metabolic patterns exist only in relation to one another, and one of the major problems of biochemistry is to understand how the patterns fit together in such a way as to be self-regulating and self-reproducing.

Many metabolic pathways are connected by a common substrate. Thus pyruvate can serve as a substrate for at least five reactions

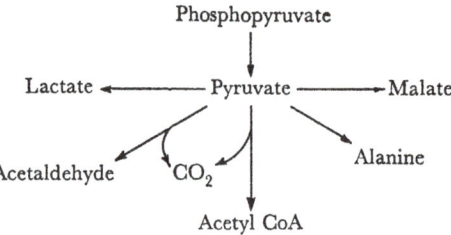

The enzymes acting on a common substrate direct that substrate into various metabolic pathways, and it is important to know the quantitative significance of the various pathways.

Consider two simultaneous first-order reactions

$$B \xleftarrow{k_b} A \xrightarrow{k_c} C;$$

rate formation of $B = V_B = k_b(a-x)$,

rate formation of $C = V_C = k_c(a-x)$,

therefore $\dfrac{V_B}{V_C} = \dfrac{k_b}{k_c}$,

where a is the initial concentration of A and x is the amount decomposed after a given time.

The fact that the ratio of the concentrations of B to C at any instant is a constant k_b/k_c and independent of time, is the basis of the method used in chemistry as a test for simultaneous side reactions. This formulation, sometimes called the Wegscheider test, is applicable only if both reactions are of the same order, and is not applicable to enzyme reactions.

Consider two simultaneous enzyme reactions

$$B \xleftarrow{E_B} A \xrightarrow{E_C} C.$$

The Michaelis equations give

$$V_B = \frac{(V_{\max}A \to B)A}{(K_m A \to B) + A}, \quad V_C = \frac{(V_{\max}A \to C)A}{(K_m A \to C) + A},$$

$$\frac{V_B}{V_C} = \frac{(V_{\max}A \to B)}{(V_{\max}A \to C)} \frac{(K_m A \to C) + A}{(K_m A \to B) + A}.$$

$V_{max} A \to B$ and $V_{max} A \to C$ represent the rate of formation of B and C when the enzymes are saturated with substrate A. $K_m A \to B$ and $K_m A \to C$ are the concentrations of A which give half maximum velocity for the reactions $A \to B$ and $A \to C$ respectively. The effect of substrate concentration on the ratio of the enzyme rates is shown in fig. 16.

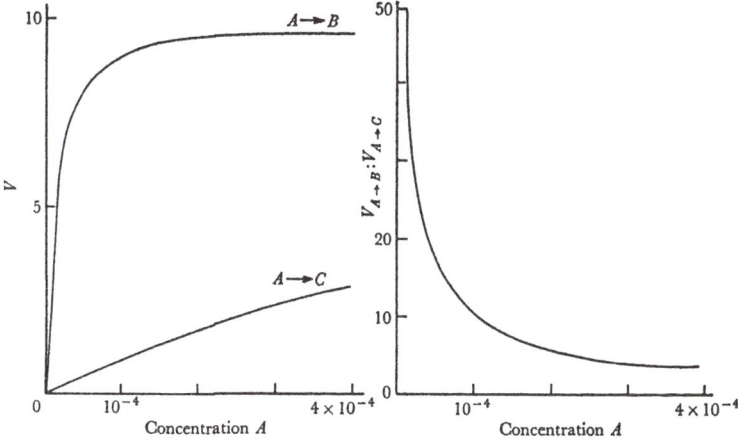

Fig. 16. Effect of substrate concentration on the velocities of simultaneous enzyme-catalysed reactions. The curves are calculated for the following conditions:

$$V_{max} A \to B = V_{max} A \to C; \quad K_m A \to B = 10^{-5} M; \quad K_m A \to C = 10^{-3} M.$$

The large effect of substrate concentration on the ratio of velocities of simultaneous reactions is of considerable practical importance. Suppose we have evidence that A is the immediate, but not necessarily the only precursor of B and C, and wish to determine the ratio of the rate of production of B and C. A possible approach would be to supply the tissue with radioactive A and after a given time to determine the amount of radioactivity in B and C. The ratio of counts in B and C is clearly a direct estimate of the ratio of the two reaction rates under the given experimental conditions. However, the feeding of A to the tissue may well increase the internal concentration of A, and if this happens the observed ratio of rates may differ markedly from the ratio in the absence of external A. An important example is the metabolism of glucose-6-phosphate which can be converted into fructose-6-phosphate and oxidised to 6-phosphogluconate. Glucose-6-phosphate is thus the branching point

23

from which carbohydrate metabolism may be directed into the Embden–Meyerhof pathway or into the pentose-phosphate pathway. If we wish to determine the relative contribution of these pathways, care must be taken not to change the concentration of glucose-6-phosphate. This point is further discussed on p. 56.

Let us now consider simultaneous reactions, one of which is showing zero-order kinetics. Feeding A to the tissue will not increase the production of the compound formed by the zero-order reaction. Consequently, if feeding A does not increase the production of B, we must avoid the conclusion that A is not the precursor of B. It should also be noted that feeding A may lead to the production of compounds not normally produced from A. The increased concentration of A may permit enzymes with a low affinity for A to convert it into a variety of compounds. This effect, sometimes known as *shunt* metabolism, is of relatively little importance in higher plants, but is of considerable economic significance in relation to the metabolism of the fungi (Foster, 1949).

CHAPTER 2

Organisation and Structure

THE organisation of metabolic patterns in free solution is made possible by the chemical specificity of enzymes, and this specificity is determined by the structure of the 'active site' of the enzyme. For example, enzyme reactions are stereospecific and the stereospecificity can be explained on the basis of a three-point attachment between the substrate and the enzyme (Ogston, 1948). The substrate molecule illustrated in fig. 17 has two identical groups (drawn in black), and for convenience labelled A and B. A non-enzymic reaction involving the replacement of this group could not distinguish between A and B and would give a racemic product. An enzyme reaction can differentiate between A and B and so give rise to an optically active product. In case I, the substrate molecule is orientated to make a three-point attachment with the enzyme (when the patterns of enzyme model and substrate model match). In case II, the substrate molecule is orientated to allow group A to attach to the enzyme, but a three-point attachment is then impossible. Thus the underlying reason for the stereospecificity of enzyme action lies in the asymmetry of the enzyme surface— a racemic-enzyme surface would give a racemic mixture of products.

The influence of structure on function is to be found at all levels of biological organisation, ranging from the transmission of genetic information by a coding system based on the nucleotide sequence in nucleic acid, to the adaptations of form which are studied by comparative anatomists.

Cellular organisation is readily observed under the low-power microscope and one of the most characteristic features of mature plant cells is the presence of a large central vacuole. Cells of higher animals are bathed in blood and waste products of metabolism are removed from the blood by the kidneys. In contrast, cells of higher plants possess a large central vacuole and so carry their aqueous environment within themselves.

Waste material is stored within the vacuole so that the elimination of waste is generally restricted to leaf-fall and the dispersal of fruits. The lack of a secretory mechanism is balanced by the fact that, with the exception of carnivorous plants and plants forming symbiotic associations with fungi or bacteria, all complex molecules are synthesised within the plant, and are not derived, as in animals, from a foreign organism.

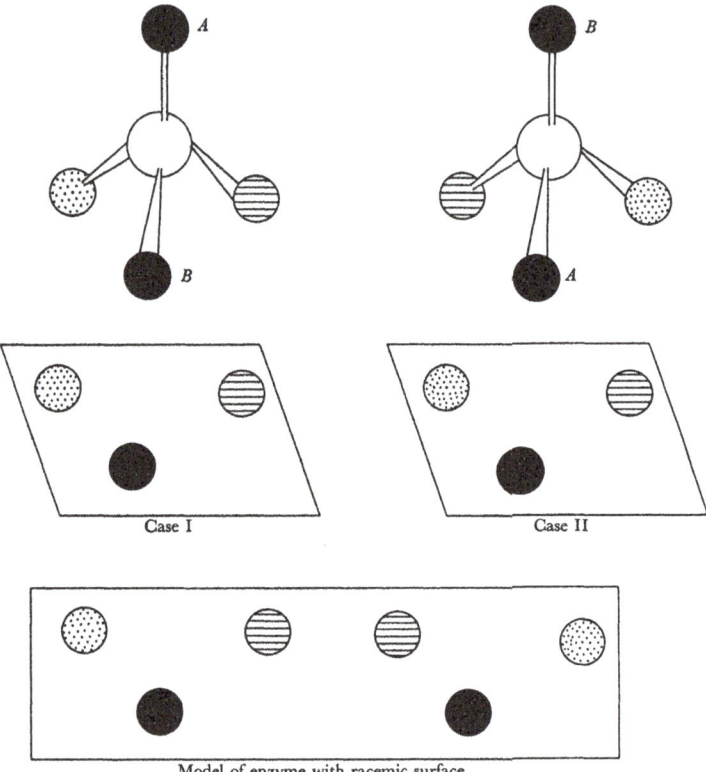

Model of enzyme with racemic surface

Fig. 17. Diagrammatic representation of optical specificity based on a three-point attachment between enzyme and substrate.

The term 'waste product of metabolism' has a fairly well-defined meaning in animal biochemistry, but its meaning is less well defined in plant biochemistry. A plant may be considered as a chemical factory, driven by light energy, the photosynthetic process being so efficient that over a wide range of environmental conditions large quantities of complex organic compounds are produced. These compounds represent the net

26

gain of synthetic over degradative reactions and, whether or not we consider a compound waste, is a matter of definition. Cellulose and lignin are not usually considered as waste products because of their participation in the anatomy of the plant, whereas many alkaloids are considered waste products because we cannot ascribe a function to them. Other compounds are more difficult to classify—for example, glycine betaine has a function in methylation but is present in some plants, for example sugar-beet, at such high concentration, that it is sometimes considered as a waste product. Similarly many plants, for example potatoes, store such large quantities of starch that it could be called waste, or at least, excess starch.

Plants accumulate many inorganic compounds, frequently within the vacuole; thus potassium is accumulated by all plant cells, barium by Brazil nuts, silica is deposited on the outer surface of many grasses, aluminium oxide may account for as much as 30–80 % of the wood ash of the Australian plant *Orites excelsa*, and fluorine is found as fluoracetic acid in the South African plant *Dichapetalum cymosum*.

The vacuole presents problems in the technique and interpretation of experiments. Thus many plants contain large quantities of organic acids and tannin which are released from the vacuole during homogenisation and may inactivate enzymes. It is fortunate for the study of plant enzymology that man has selected and cultivated many food plants containing little tannin and with vacuolar sap whose pH is near neutrality. As an example of the complications which the vacuole produces in feeding experiments, consider the metabolic system

$$A \rightarrow B \rightarrow C.$$

In a homogeneous system, the feeding of ^{14}C-labelled A would produce labelled B and C and the specific activity of B (defined as the number of disintegrations per minute per μmole of the compound) must be greater than the specific activity of C. If, however, B is present in the vacuole and the reaction $B \rightarrow C$ takes place in the cytoplasm, the specific activity of C may appear greater than that of B because, during isolation, B stored in the vacuole may dilute labelled B present in the cytoplasm. Consequently the analysis of feeding experiments usually requires mathematical analysis of the experimental data (see Aronoff, 1956).

27

The view that the tonoplast separates the vacuolar contents and the cytoplasm is supported by the demonstration (Bennet-Clark and Bexon, 1943) that vacuolar sap supplied externally to beetroot disks produces a large increase in the rate of respiration. This point has not been sufficiently appreciated, and Bonner (1950) has suggested that if a plant accumulates a particular compound the enzyme attacking that compound must be absent or present in low concentration. Stekol (1958) has argued that because bean stems contain large quantities of malonic acid (a competitive inhibitor of succinic dehydrogenase), the oxidation of succinate cannot proceed via succinic dehydrogenase.

Present-day knowledge of subcellular organisation is largely due to the simultaneous development of the electron microscope and methods of differential centrifugation. Electron microscopy has revealed the detailed structure of nuclei, chloroplasts and mitochondria and demonstrated that the optically clear ground substance of cytoplasm has an extensive system of tubules or vesicles, limited by a thin membrane and interconnected in a more or less continuous network—the endoplasmic reticulum. Outside, and frequently attached to the endoplasmic reticulum, are small dense particles about 15 mμ in diameter.

Some of the structures discernible under the electron microscope may be isolated from homogenates by differential centrifugation, but such preparations are seldom free from other cell components. Particulate material isolated from homogenates prepared in buffered sugar solutions by centrifugation at 20,000 g has been shown to carry out all the reactions of the Krebs cycle (Millerd, Bonner, Axelrod and Bandurski, 1951; Davies, 1953). Such material was at first rather loosely identified with mitochondria, but subsequent work has shown that the identification was substantially correct.

Microsomes are too small to be observed under a light microscope and were at first recognised in animal material as 'phospholipide-ribonucleoprotein complexes' which could be isolated from homogenates prepared in dilute phosphate buffer by centrifuging at 18,000g for 1hr (Claude, 1940). More recently, isotonic sugar solutions have been used to prepare homogenates and centrifugal forces of the order 100,000 g are then necessary to sediment the microsomes. Microsomes isolated in this manner from the white petiole of silver-beet

(*Beta vulgaris*) contain a DPNH cytochrome *c* reductase system, which is not inhibited by antimycin A, whilst the reductase activity of mitochondria is sensitive to very low concentrations of the antibiotic. The view that the reductase activity of microsomes is distinct from that of mitochondria is supported by the finding that the reductase activity of microsomes is intimately associated with cytochrome b_3 which is exclusively localised in the microsomes (Martin and Morton, 1956). Microsomes isolated from peanut cotyledons contain an enzyme system which catalyses the α-oxidation of fatty acids, in contrast to the fatty acid oxidation system of mitochondria which effects the β-oxidation of CoA derivatives of fatty acids (Stumpf and Bradbeer, 1959). The most important function of microsomes is their participation in protein synthesis (Webster, 1959). These biochemical properties are associated with a particular cell fraction obtained by differential centrifugation. Attempts to identify microsomes with a structure discernible under the electron microscope has led to some confusion of terms. Some workers consider that the microsomal fraction represents vesicles derived from the endoplasmic reticulum and associated particles rich in ribosenucleic acid, whereas other workers equate the terms microsome and ribonucleoprotein particle. At a recent symposium (Roberts, 1958) it was suggested that the term *ribosome* be adopted to designate ribonucleoprotein particles in the size range 20–100 Svedberg units. (The rate of sedimentation in a centrifuge is usually measured in terms of a sedimentation coefficient S, which is the velocity for unit centrifugal force and has the dimensions of time. A sedimentation coefficient of 10^{-13} s is taken as one Svedberg unit and data obtained in the ultra-centrifuge is usually given for S_{20}, since sedimentation is usually measured at this temperature.)

Ribosomes isolated from pea seedlings (Ts'o, Bonner and Vinograd, 1958) contain 35–40 % ribosenucleic acid by weight, have a molecular weight of 4–4·5 million and appear spherical under the electron microscope. Molecular weight in this context is the microsomal gram molecular particle weight and is numerically equal to the weight of a ribosome multiplied by Avogadro's number. Ribosomes appear to consist of six nucleoprotein subunits, held together by bonds from calcium and magnesium to ribosenucleic acid and protein. The molecular weight of the subunit is about 7×10^5 and on the basis of a

29

ribosenucleic acid content of 40 %, the amount of ribosenucleic acid present would correspond to a single molecule of molecular weight approximately $2 \cdot 8 \times 10^5$. Such a molecule would contain less than 1000 nucleotides and on the basis of present coding theories, this could code a sequence of some 300 amino acids (see Bonner, 1959).

The origin of the ribosome is uncertain, but it may be synthesised *de novo* in the nucleus and pass into the cytoplasm through pores in the nuclear membrane. It is tempting to speculate that in those cells possessing an endoplasmic reticulum, the ribosomes pass through the tubules of the reticulum which in some electron micrographs appears to be continuous with the nuclear membrane.

Perhaps the most interesting property of ribosomes is their association with protein synthesis (Bonner, 1959; Webster, 1959). Amino acids are activated and bound to the activating enzyme which is present in the soluble fraction of the cell. The reaction may be represented as

$$\text{Enz} + \text{amino acid} + \text{ATP} \rightleftharpoons \text{enz} - \text{amino acid} - \text{AMP} + \text{PP}.$$

This reaction has been detected by means of the exchange reaction between ^{32}P-labelled pyrophosphate and the terminal phosphate of ATP which takes place in the presence of an amino acid. The reaction has also been studied by observing the transfer of activated amino acid to hydroxylamine

$$\text{Enz} - \text{amino acid} - \text{AMP} + \text{hydroxylamine} \rightleftharpoons \text{enz}$$
$$+ \text{AMP} + \text{amino acid hydroxamate}.$$

It is probable that specific activating enzymes exist for each amino acid. In the cell the activated amino acid is transferred to an acceptor which is a soluble polynucleotide

$$\text{Enz} - \text{amino acid} - \text{AMP} + \text{polynucleotide} \rightleftharpoons \text{enz}$$
$$+ \text{AMP} + \text{amino acid} - \text{polynucleotide}.$$

The exact nature of the polynucleotide is unknown and there may well be separate polynucleotide acceptors for each amino acid. The next step in the synthesis of protein is the transfer of the amino-acid polynucleotide to the surface of the ribosome, where it is incorporated into a protein associated with the ribosome.

The interpretation of these observations is that the ribosome links the activated amino acids into a peptide chain and, though there is no direct experimental evidence, it is reasonable to assume that the ribosome acts as a template to determine the sequence of amino acids.

Ribosomes are too small for any details of their structure to be discernible under the electron microscope, but mitochondria are large enough to be sectioned and a great deal of their structure can be studied with the electron microscope and related to biochemical studies on isolated mitochondria. This biochemical investigation has shown that mitochondria can catalyse all the reactions of the Krebs cycle and, after disintegration of the particles, some enzymes such as fumarase are found to be soluble whilst other enzymes such as succinic dehydrogenase are associated with insoluble material. Electron microscopy has shown that mitochondria are limited by double membranes, the inner membrane being folded so that it forms a series of projections into the centre of the organelle (Heitz, 1957). A reasonable interpretation is that the membranes ensure that all the enzymes of the Krebs cycle are retained within the organelle, some of the enzymes being located in or on the membranes, whilst others are in solution.

An obvious consequence of the retention of enzymes within an organelle is to reduce the distance between enzyme molecules and thereby reduce the time necessary for substrates and coenzymes to diffuse from enzyme to enzyme. Dixon and Webb (1958) point out that all that is necessary for the efficient working of a multienzyme system is that the components should be held sufficiently near together to make the transit time small, either by confining them within a membrane, by adsorption on a surface, by association under the influence of electrostatic or other forces, or by combination with a colloidal particle. Green (1957), on the other hand, has developed the idea that enzymes are rigidly arranged in a special sequence, and maintains that the coenzymes of the pyridine nucleotide dehydrogenases are bound to the enzymes within the mitochondria and are not free to diffuse as required by the views proposed by Dixon and Webb. Perhaps the best example supporting the views of Green comes from the electron transporting system associated with fatty acid oxidation (Beinert and Crane, 1956). The first step in the oxidation of a fatty acyl-CoA substrate is catalysed by a metalloflavoprotein, with the

31

reduction of the flavin and the formation of a stable compound between α, β-unsaturated fatty acyl-CoA and the reduced flavoprotein. The α, β-unsaturated fatty acyl-CoA cannot be acted upon by a further enzyme until it is released from the enzyme surface, and this can only occur after the reduced flavin has been oxidised by another flavoprotein called the electron-transferring flavoprotein (ETF). The flavin nucleotides are unable to diffuse from one enzyme to the other so that the cycle of oxidation and reduction would seem to require the adjacent positioning of the enzymes.

Structural integrity appears essential for respiratory chain phosphorylation and in the theory of oxidative phosphorylation proposed by Grabe (1958) (see p. 45) the correct spatial arrangement of the reactants is assumed. Finally, the possibility that the layered membrane of mitochondria and similar layered structures in chloroplasts and nuclear membranes may be connected with a radiationless transfer of energy has been raised by Wald (1956) and discussed at a recent meeting of the Faraday Society.

In the previous chapter it was argued that because a number of the enzymes of the pentose-phosphate pathway have rather wide specificities, the reactions of the pathway can only occur in the form of a cycle, if the enzymes themselves are arranged in an ordered sequence. That the enzymes of this pathway may be physically associated is suggested by the finding that the entire group of enzymes show a similar behaviour on centrifugation. Thus, pentose-phosphate cycle activity remains in the supernatant after centrifugation of a liver homogenate at $105,000g$ for 2 h, but is associated with a red layer which sediments after centrifugation at $144,000g$ for 16 h (Newburgh and Cheldelin, 1956). The existence of a multienzyme unit in which the individual enzymes are physically associated has been raised from time to time. Schneider and Hogeboom (1951) found that analysis of soluble mitochondrial protein in the ultracentrifuge indicated the presence of only three proteins which led them to discuss the existence of a 'giant' molecule possessing many catalytic sites with different enzyme activities. Whilst such a complex offers a number of biological advantages, the experimental basis is not very firm, since a complex mixture of enzymes may give a single peak in the ultracentrifuge. An extract of muscle shows eight components after electrophoretic analysis, but only two can be detected in the analytical ultracentrifuge. Similarly the cytoplasmic proteins of leaves show two components: Fraction I which is homogeneous and has been tentatively identified with the carboxylation enzyme of photosynthesis (Dooner, Kahn and Wildman, 1957) and Fraction II which gives a single peak, but this peak is asymmetrical and electrophoresis indicates a heterogeneous mixture of proteins.

The concept of a close association between enzymes has recently been developed by Green (1957), who points out that

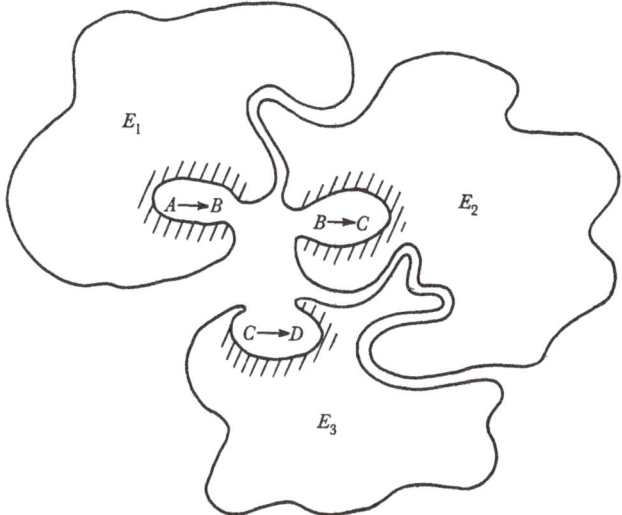

Fig. 18. Diagrammatic representation of association between enzymes. The enzymes catalysing the sequence $A \to B \to C \to D$ are arranged to reduce the diffusion time of the intermediates in the sequence. Shaded areas represent 'active sites' on the enzymes.

the separation of two enzymes which follow one another in the fatty acid oxidation system (p. 58) frequently presents great difficulty. Four enzymes participating consecutively in fatty acid oxidation are

(1) an acyl dehydrogenase,
(2) a hydrase,
(3) a hydroxyacid dehydrogenase,
(4) a cleavage enzyme.

During the purification of these enzymes it was found that (2) was a persistent contaminant of (3) and (3) was a contaminant of (4), but (2) was not a contaminant of (4). Green proposes that there are specific enzyme-enzyme interactions, which are specific and allow a large number of enzymes to be linked in an intricate pattern (fig. 18).

With reference to plants, the functional association of malic dehydrogenase and glutamic-oxaloacetic transaminase is suggested by the apparently simultaneous labelling of malate and aspartate during the dark fixation of carbon dioxide. The physical association of these two enzymes throughout a 200-fold purification has been observed in the author's laboratory (Davies and Ellis, 1959). With reference to the reactions of the pentose-phosphate pathway, pentose-phosphate isomerase is associated with transketolase over a 150-fold purification of spinach transketolase (Horecker, Smyrniotis and Hurwitz, 1956) and xylulose-5-phosphate epimerase is a persistent contaminant of crystallised yeast transketolase (De la Haba, Leder and Racker, 1955).

The possibility of a physical and functional association between enzymes is worthy of further investigation which is being made possible by the availability of preparative ultracentrifuges in a number of laboratories.

CHAPTER 3

Bioenergetics

THE most useful concept of thermodynamics which can be applied to the study of metabolism is that of free energy. Four equations relating to the change of free energy in a system such as

$$A + B \rightleftharpoons C + D$$

are given below:

$$\Delta G = \Delta H - T\Delta S, \qquad (1)$$

$$\Delta G^0 = -RT\ln K, \qquad (2)$$

$$\Delta G = \Delta G^0 + RT\ln \frac{(C)(D)}{(A)(B)}, \qquad (3)$$

$$\Delta G^0 = -nFE_o. \qquad (4)$$

The change in free energy (ΔG)[1] of a reaction is the amount of energy available for useful work. ΔH is the amount of heat released by the reaction, T is the absolute temperature and ΔS is the change in entropy of the system. The concept of entropy can be illustrated by reference to an ice–water mixture. On supplying heat to the mixture, some ice will melt, but the temperature will remain constant. All forms of energy have two factors, an intensity factor and a capacity factor, thus the intensity factor of electrical energy is voltage and the capacity factor is charge. When heat is absorbed by the ice–water mixture, the intensity factor—temperature—remains constant, consequently there must be an increase in the capacity factor—entropy. The increase in entropy when ice melts is associated with the greater freedom of motion of the water molecules. The water molecules are more random and this is, therefore, a more probable state than the ordered arrangement in ice. Thus an increase in entropy corresponds to an increase in randomness and probability. The complex molecules of living organisms have less freedom and are less random than their constituent

---

[1] In British literature, Gibbs free energy (G) is related to Helmholtz free energy (F) by the expression $G = F + PV$. In American literature Gibbs free energy is often given the symbol F.

parts, thus the synthesis of complex molecules such as sugars and proteins involves a decrease in entropy. This has been succinctly stated by Schroedinger in the now famous quotation 'living organisms feed on negative entropy'. It is perhaps useful when thinking about entropy to have a molecular model. Molecules have two main forms of energy: one form holds the atoms together and it is this energy which can provide useful work; the other form of molecular energy is due to random intramolecular vibrations, rotations and translations and corresponds to entropy. The relationship between entropy and probability is due to the fact that the vibrational and rotational motions have quantised energy levels and the larger the number of energy levels available to a molecule the more probable is that particular molecular structure and the greater is its entropy.

Equation (1) can be used to determine the free energy of a reaction. ΔH can be determined by calorimetry, and the entropy of each reactant can be determined from accurate specific heat data obtained at temperatures as near absolute zero as possible.

Equation (2) states the relationship between the standard free energy change (ΔG^0) and the equilibrium constant K. The standard conditions are taken as $25°$ C and in the case of gases a pressure of 1 atmosphere, for solutes 1·0 molal activity, whilst the standard for water is taken as unity rather than 55·5 which is the molar activity of water for dilute solutions. It should be noted that $\Delta G'$ is frequently used in biochemistry and is identical with ΔG^0, except that the standard condition of H^+ ion is that of the pH specified instead of 1 molal activity, that is, pH o.

Equation (3) states the relationship between ΔG and concentration. If the reaction is at equilibrium

$$RT\ln\frac{(C)\,(D)}{(A)\,(B)} = -G^0$$

and consequently ΔG is zero. Equation (3) is important because it enables one to calculate the free-energy change for any particular condition, provided the standard free-energy change is known.

Equation (4) gives the relationship between ΔG^0 and the oxidation-reduction potential E_o. F is the Faraday constant

36

(23,068 cal per volt equivalent) and n is the number of electrons involved. It should be noted that there are two conventions for recording E values. In chemistry the more easily *oxidised* compounds are given positive values and the equation for the half cell is

$$E_h = E_o - \frac{RT}{nF} \ln \frac{\text{Oxidant}}{\text{Reductant}}.$$

In biochemistry positive values of E are given to the more easily *reduced* compounds:

$$E_h = E_o + \frac{RT}{nF} \ln \frac{\text{Oxidant}}{\text{Reductant}}.$$

The usefulness of free energy data lies in the fact that they enable us to decide if a particular reaction is possible. If ΔG is

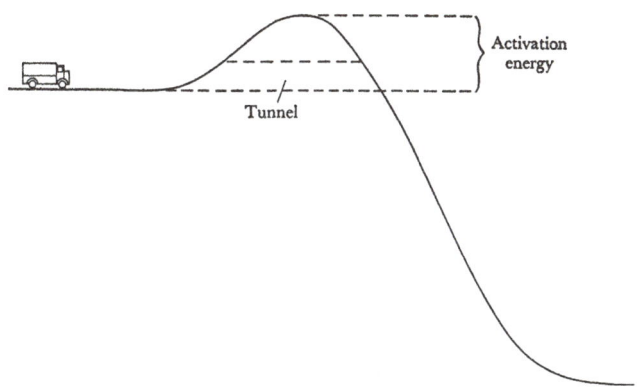

Fig. 19. Analogy between positional energy and chemical energy.

positive the reaction is *not* spontaneous, but if ΔG is negative then the reaction can occur spontaneously, but it does not follow that it *will* take place. For example, this page consists mainly of cellulose, and the reaction of cellulose with oxygen has a large negative change in free energy. The paper, however, is perfectly stable until sufficient energy is supplied to make it unstable. If a lighted match is used to give 'activation energy' the paper will burn to carbon dioxide and water with a release of heat energy. The situation can be likened to a car near the top of a hill (see fig. 19). The car is perfectly stable, but if it is pushed up the incline (given activation energy) it becomes unstable and runs down the hill with a loss of gravitational energy.

In an enzyme-catalysed reaction the enzyme acts by lowering the activation energy. In terms of the hillside analogy, the action of an enzyme can be compared to a tunnel cut through the incline.

When discussing bioenergetics we must take notice of the well-known comparison between respiration and burning. The well-known saying of Rosenfeld that 'fat burns in the fire of carbohydrate' may have some poetical justification, but it should be recognised that the analogy drawn between respiration and burning can be misleading. The plant is not a heat engine and cannot convert temperature and pressure changes into useful work. When a compound such as fat is burned, its chemical energy is released and appears as heat energy, but when a fat is oxidised by a plant, there is a relatively small production of heat and most of the chemical energy is conserved. Thus in the germination of castor bean (*Ricinus*) seeds there is a loss of fat but an increase in the amount of carbohydrate. It is sometimes said that the energy derived from oxidation is used to *drive* synthetic or anabolic reactions. This concept of one reaction driving another is best considered in terms of an equilibrium between reactions which share common intermediates. Plants make use of two types of compounds to couple catabolic and anabolic reactions—reduced coenzymes such as DPNH and TPNH and anhydrides such as ATP, thiol esters of CoA and formyl-THFA.

Fats and carbohydrates are highly reduced compounds which can, by means of the appropriate enzymes, effect the reduction of DPN and TPN. The reduced coenzymes are then available to participate in synthetic reactions. For example, glucose-6-phosphate may reduce TPN and the TPNH may be oxidised in the fixation of carbon dioxide by malic enzyme.

$$\text{Glucose-6-phosphate} + TPN^+ \rightleftharpoons \text{6-phosphogluconate}$$
$$+ TPN + H^+, \quad \Delta G = -6 \cdot 5 \text{ kcal.}$$
$$H^+ + TPNH + \text{pyruvate} + CO_2 \rightleftharpoons \text{malate} + TPN^+,$$
$$\Delta G = +2 \cdot 0 \text{ kcal.}$$

Since the second reaction is endergonic, we say that it requires energy to *drive* it, the energy being provided by the first reaction which is exergonic. It is less ambiguous to say that the equilibrium of the first reaction favours the production of TPNH and it is the high ratio TPNH/TPN which by mass action *drives* the second reaction in the direction of carbon dioxide fixation.

If the transfer of energy from catabolism to anabolism is to be reasonably efficient, then heat losses in the enzyme reactions leading to the synthesis of the intermediates common to the catabolic and anabolic reactions must be small. When a closed system is at equilibrium there is no change in free energy and the rate of entropy production is zero. In an open system in the steady state, the rate of entropy production is minimum (Prigogine, 1955), and the free energy change is minimum when the concentrations of reactants are maintained close to equilibrium. Thus, in a multi-enzyme system, heat losses are minimal when the individual reactions are fast compared with the overall rate, so that the intermediates are maintained close to their equilibrium values. Mitochondria contain the enzymes of the Krebs cycle and the activities of fumarase and aconitase are high compared to the overall rate, so that there should be little heat loss at these two steps (see table 5).

TABLE 5. *Activity of soluble enzymes from pea mitochondria* (*activities at pH 7·4, measured as μmoles product/hr/mg protein*)

Enzyme	Activity
Glutamic dehydrogenase	0·48
Alcohol dehydrogenase	0·20
Adenylate kinase	8·8
DPN cytochrome *c* reductase	600
Aconitase	550
Fumarase	3000

A wide range of anhydrides participate in energy transfer, but the most important is adenosinetriphosphate (ATP). Lipmann (1941) introduced the useful concept of the 'high-energy bond' and in particular the high-energy phosphate bond represented as $\sim$ P. The free energy of hydrolysis of an ester phosphate under standard conditions is of the order 2–5 kcal, but the free energy of hydrolysis of the terminal bond of ATP is of the order 7–8 kcal. The use of the term high-energy bond has been criticised by Gillespie, Maw and Vernon (1953) because it conflicts with chemical usage. Bond energy, sometimes called the heat of formation of the bond, is the average amount of energy required to dissociate bonds of the same type in one mole of a given compound (table 6).

39

TABLE 6. *Bond energies*

Bond	C—C	C—O	C=O	C—N	C=N	C—H
Energy (kcal)	57	70	150	49	94	87

The term high-energy bond as used by Lipmann should, therefore, be replaced, and alternative terms such as group-transfer potential and high-energy compound have been suggested.

There are four types of organic phosphate compounds which have high free energies of hydrolysis:

1. Carboxyl phosphate:

$$\begin{array}{ccc} & O & O \\ & \| & \| \\ R-&C-O-P-&OH \\ & & | \\ & & OH \end{array}$$

2. Enol phosphates:

$$\begin{array}{cc} | & \\ -C & O \\ \| & \| \\ C-O-P-&OH \\ | & | \\ & OH \end{array}$$

3. Pyrophosphates:

$$\begin{array}{ccc} O & & O \\ \| & & \| \\ R-P-&O-&P-OH \\ | & & | \\ OH & & OH \end{array}$$

4. Amine phosphates:

$$\begin{array}{cc} NH & O \\ \diagdown & \| \\ C-NH-&P-OH \\ | & | \\ R & OH \end{array}$$

The reasons why these particular compounds should be high-energy compounds have been discussed by Oesper (1950) and more recently by Grabe (1958). The main idea is that of 'opposing resonances'. The moieties which are formed on hydrolysis can be considered as competing in their resonance for the electrons of the bridge atom —O— or —H— and consequently the combined form has less resonance and is, therefore, less stable than the hydrolysed system. Oesper has pointed out that there are twenty-nine resonance forms of the phosphate ion HPO_4^{2-}, for example,

$$\begin{array}{cccc} O^- & O^- & O^- & O^- \\ | & | & | & | \\ HO-P^+-O^- & HO-P=O & HO^+=P-O^- & HO-P^+=O \\ | & | & | & | \\ O^- & O^- & O^- & O^{..} \end{array}$$

and the carboxyl group has two canonical forms:

$$\begin{array}{cc} \overset{OH}{\underset{}{R-C}} & \overset{O^+-H}{\underset{}{R-C}} \\ \diagdown O & \diagdown O \end{array}$$

Thus there are fifty-eight conceivable forms of the carboxyl phosphate, but thirteen of these forms are not possible, so that the carboxyl phosphate group has fewer resonance forms and is thus unstable.

The electron distribution in the carboxyl-phosphate group has been calculated by Grabe (fig. 20), and this distribution enables us to consider the instability of the carboxyl-phosphate

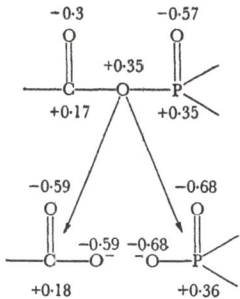

Fig. 20. Electron distribution in the carboxyl-phosphate group. The numbers are units of electronic charge due to the two π electrons of the C=O bond, the lone pair of the central O atom and the electrons of the phosphoryl group (PO$_3$).

in terms of two opposing forces. One force tends to produce the electron distribution of the carboxyl ion, which represents a low-energy level, and in so doing pulls electrons from the P—O bond tending to produce an activated phosphoryl group. The other force tends to produce the electron distribution of the phosphate ion by pulling electrons from the C—O bond and thus tends to produce an activated keto group.

Experiments with ^{18}O have shown that the carboxyl-phosphate group can be cleaved on either side of the —O— bridge. Thus, for example, in the reduction of 1:3-diphospho-glycerate the carbon oxygen bond is broken

$$\begin{array}{ccc} & & \\ R-\overset{O}{\underset{}{C}}\text{---}O\text{---}\overset{O}{\underset{O^-}{P}}\text{---}OH & \rightleftharpoons & R-\overset{O}{\underset{}{C}}\text{---}H + O\text{---}\overset{O}{\underset{O^-}{P}}\text{---}OH \\ +\text{DPNH}+\text{H}^+ & & +\text{DPN}^+ \end{array}$$

41

whereas in the reaction catalysed by phosphoglyceryl kinase

$$\underset{R}{\overset{O}{\parallel}}C\!-\!O\!-\!\underset{O^-}{\overset{O}{\underset{\parallel}{\overset{\parallel}{P}}}}\!-\!OH + ADP \rightleftharpoons R\!-\!\overset{O}{\overset{\parallel}{C}}\!-\!OH + ATP$$

the O—P bond is broken.

The point of cleavage is determined by a specific enzyme which, therefore, determines the direction of electron displacement and the activation of the keto or phosphoryl groups. If the P—O bond is split the phosphoryl group is activated and if the C—O bond is split the keto group is activated and can take part in anabolic reactions. Similarly the point of cleavage of ATP determines which group shall be activated (fig. 21). For

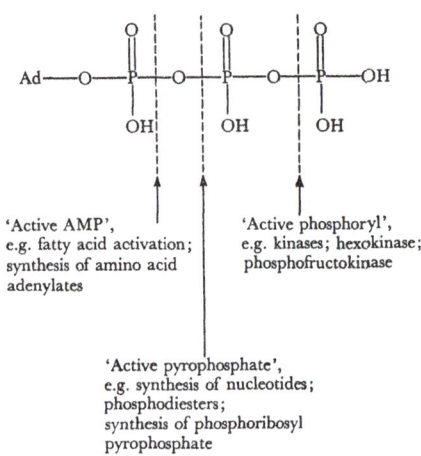

'Active AMP',
e.g. fatty acid activation;
synthesis of amino acid
adenylates

'Active phosphoryl',
e.g. kinases; hexokinase;
phosphofructokinase

'Active pyrophosphate',
e.g. synthesis of nucleotides;
phosphodiesters;
synthesis of phosphoribosyl
pyrophosphate

Fig. 21. Points of cleavage of ATP.

example, the activation of acetate according to the overall equation

$$\text{Acetate} + \text{ATP} + \text{CoA} \rightleftharpoons \text{acetyl CoA} + \text{AMP} + \text{PP}$$

proceeds by the following steps:

42

$$\text{Ad}-\text{O}-\overset{\displaystyle\text{O}}{\underset{\displaystyle\text{OH}}{\text{P}}}-\overset{18}{\text{O}}-\overset{\displaystyle\text{O}}{\text{C}}-\text{CH}_3 \quad \longrightarrow \quad \text{Ad}-\text{O}-\overset{\displaystyle\text{O}}{\text{P}}-\overset{18}{\text{OH}}$$

$$+\,\text{CoASH} \qquad\qquad +\quad \text{CH}_3-\overset{\displaystyle\text{O}}{\text{C}}-\text{S}-\text{CoA}$$

The enzymic cleavage of ATP produces an activated AMP moiety which combines with acetate to give adenyl acetate. The enzymic splitting of adenyl acetate at the C—O bond produces an activated keto group which combines with CoA to give acetyl CoA.

Under anaerobic conditions ATP is produced by the reactions of glycolysis (fig. 4). The oxidation of an aldehyde to the corresponding acid is a highly exergonic reaction. For example, the TPN specific triosephosphate dehydrogenase found in leaves by Arnon, Rosenberg and Whatley (1954) does not require phosphate and catalyses the irreversible reaction

$$
\begin{array}{l}
\text{CHO} \\
|\\
\text{CHOH} \\
|\\
\text{CH}_2\text{O}\circledP
\end{array}
+\,\text{TPN}^+ + \text{H}_2\text{O} \rightarrow
\begin{array}{l}
\text{COOH} \\
|\\
\text{CHOH} \\
|\\
\text{CH}_2\text{O}\circledP
\end{array}
+\,\text{TPNH} + \text{H}^+
$$

Leaves contain two other triosephosphate dehydrogenases, one specific for DPN, the other for TPN, but both require inorganic phosphate. The reactions catalysed by these enzymes are reversible and the product, 1:3-diphosphoglycerate, is a high-energy compound. The reaction mechanism of the DPN specific enzyme from yeast and from animal tissues has been intensively studied, and a possible reaction mechanism is depicted in fig. 22. The other reaction of glycolysis which produces a high-energy compound is the dehydration of 2-phosphoglyceric acid

$$
\begin{array}{l}
\text{CH}_2\text{OH} \\
|\\
\text{CHO}\circledP \\
|\\
\text{COOH}
\end{array}
\underset{-\text{H}_2\text{O}}{\rightleftharpoons}
\begin{array}{l}
\text{CH}_2 \\
\|\\
\text{C}-\text{O}-\circledP \\
|\\
\text{COOH}
\end{array}
$$

Here the dehydration produces an *enol*-phosphate, which has a large energy of hydrolysis because the enol-form of pyruvate is unstable relative to the carbonyl form. The production of ATP

43

by the reactions of glycolysis is an inefficient process, since only two moles of ATP are produced per mole of glucose. Under aerobic conditions pyruvate is completely oxidised via the Krebs cycle, and experiments with mitochondria have shown that during the oxidation of a number of compounds (mainly

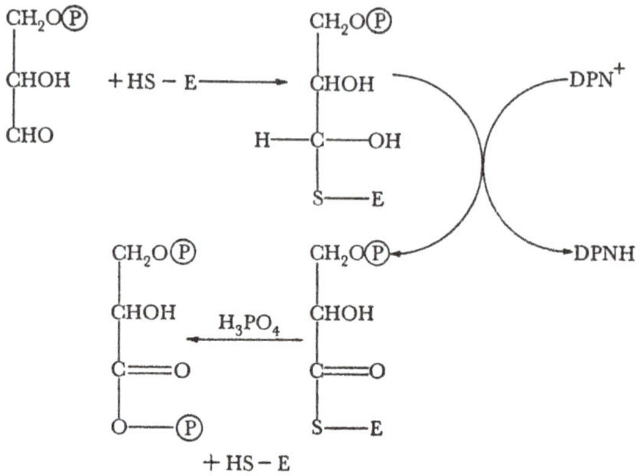

Fig. 22. Possible reaction mechanism of triosephosphate dehydrogenase.

acids of the Krebs cycle) inorganic phosphate becomes esterified. The ratio

$$\frac{\text{equivalents of inorganic phosphate esterified}}{\text{atoms of oxygen consumed}},$$

usually written as the P/O ratio, has been determined for a number of substrates with mitochondria from a variety of plants (table 7). To determine the P/O ratio, freshly isolated mitochondria are incubated with a substrate, inorganic phosphate, Mg^{2+} ions, a catalytic quantity of ADP, glucose and purified hexokinase, and the oxygen uptake determined by manometry. During the oxidation of the substrate inorganic phosphate is transferred to ADP to form ATP. The glucose and

TABLE 7. P/O *ratios obtained with mitochondria isolated from various plants*

Substrate	Mung bean	Cauliflower	Lupin	Pea leaves
Citrate	0·81	—	—	2·51
α-Oxoglutarate	1·03	2·36	3·0	2·92
Succinate	0·93	1·34	2·0	1·58
Malate	0·89	1·59	—	—

44

hexokinase function as a trap to remove the terminal phosphate of ATP and regenerate the phosphate acceptor—ADP:

$$\text{Glucose} + \text{ATP} \underset{\text{hexokinase}}{\rightleftharpoons} \text{glucose-6-phosphate} + \text{ADP}.$$

The amount of phosphate esterified may be determined by measuring either the disappearance of inorganic phosphate or the formation of glucose-6-phosphate.

The first experiments of Millerd *et al.* (1951) gave P/O ratios of approximately 1, but with subsequent improvements in technique, higher values have been obtained (Table 7). The experimental values are considered to be minimum values and on the assumption that oxidative phosphorylation involves simple stoichiometry it has been suggested that the theoretical P/O ratio for succinate oxidation is 2, for α-oxoglutarate 4, and for the remaining oxidations of the Krebs cycle 3. These ratios can be interpreted by reference to the steps in the electron-transporting system (fig. 23). There is good evidence that fig. 23 represents the overall pattern of the mitochondrial electron-transporting system, though other compounds may also be involved in the passage of electrons. There is some uncertainty, however, about the values of the oxidation reduction potentials. Thus the potentials of cytochromes a and b have been evaluated from the visual observations of Ball (1938), but a redetermination for cytochrome b by Hill (1954) has given a value which is 40 mV more positive. Within the limits imposed by these uncertainties, we note that there is sufficient energy between DPNH and flavoprotein, between cytochrome b and c and between cytochrome c and oxygen to permit the phosphorylation of at least one mole of ADP in each case. The understanding of the process of oxidative phosphorylation has been hindered by failure to demonstrate oxidative phosphorylation in a soluble system. Nevertheless, a number of theories has recently been presented and, as an example, we shall discuss the theory proposed by Grabe (1958) to account for oxidative phosphorylation in the DPNH-FAD region of the respiratory chain. Grabe's theory recognises the experimental finding that oxidative phosphorylation is intimately connected with structure. Grabe postulates that the pyridine ring of DPNH and the riboflavin ring of FAD are aligned in parallel planes so positioned that inorganic phosphate, which is hydrogen bonded to

45

the amide group of DPNH, lies between the coenzymes, and the pyridine N-atom lies opposite the C-2 atom of the riboflavin ring. The transfer of a hydrogen atom from C-4 of the pyridine ring to the N-10 position of the riboflavin, and the simultaneous transfer of an electron from the pyridine N-atom to the riboflavin, leaves the pyridine N-atom with a positive charge and

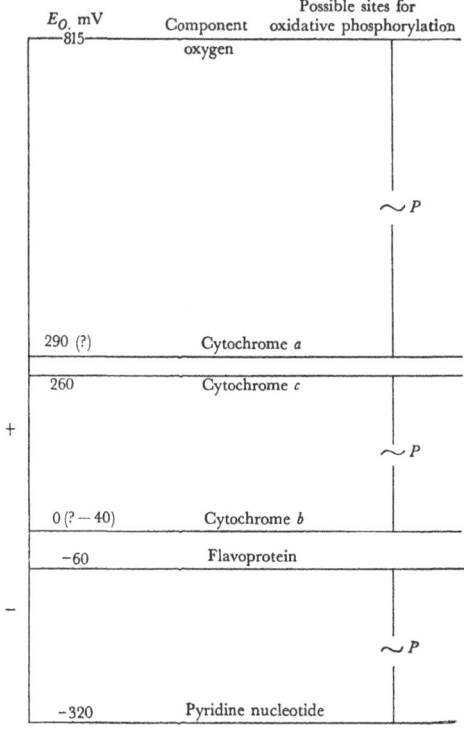

Fig. 23. Redox potentials of some components of the electron-transporting system and possible sites of oxidative phosphorylation.

gives the FAD an increased electron density. The positive charge on the pyridine N-atom attracts the negatively charged O-atom of the phosphate ion, while the electron transferred to a π-orbital of FAD increases the electron density of the CO(2) group so that the electron of the π-orbital is transferred to an orbit extending over the P-atom of the phosphate ion. This tendency to form a C—O—P bond, together with the electrostatic attraction between the pyridine N-atom and the negatively charged O-atom of the phosphate ion, strains the P—O

46

bond (I in fig. 24) which may break and so produce a phosphoryl-FADH. The splitting of the inorganic P—O bond is in accordance with the experimental demonstration (Cohn, 1949) that during oxidative phosphorylation the oxygen atoms of inorganic phosphate exchange with the oxygen atoms of water.

The phosphoryl-FADH is a high-energy compound, and during the subsequent oxidation the activated phosphoryl group is transferred to ATP.

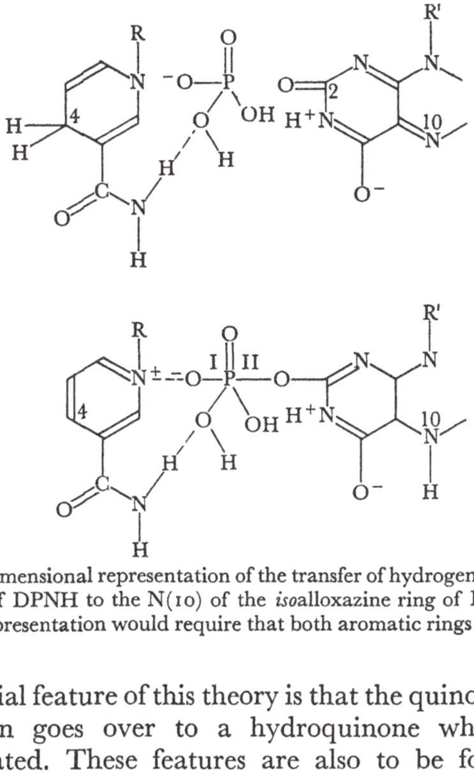

Fig. 24. Two-dimensional representation of the transfer of hydrogen from C-4 of the pyridine ring of DPNH to the N(10) of the *iso*alloxazine ring of FAD. (A three-dimensional representation would require that both aromatic rings were in parallel planes.)

An essential feature of this theory is that the quinone structure of riboflavin goes over to a hydroquinone which is then phosphorylated. These features are also to be found in the theory of oxidative phosphorylation proposed by Harrison (1958). It is thus of interest that a lipid soluble quinone, $Q275$, has been shown (Crane and Lester, 1958) to be an integral part of the electron-transporting system of beef heart mitochondria. The quinone $Q275$, so called because of its absorption peak at 275 mμ, has been isolated from potatoes and sweet potatoes. A similar quinone $Q254$ has been isolated from cauliflower buds and spinach leaves. Both quinones can be shown to be reduced to hydroquinones by succinate and beef heart mitochondria, and

47

the hydroquinones can be oxidised by mitochondria. Under similar conditions, tocopherol quinone, vitamin K and menadione have no activity (Crane and Lester, 1958).

The view that oxidative phosphorylation shows simple stoichiometry is the basis for theories involving the transfer of energy by means of coupled reactions. However, the view that the P/O ratio is a whole number is an assumption for which little direct evidence exists and P/O ratios greater than 2 for succinate have been reported (Whittam, Bartley and Weber, 1955). The validity of determining the P/O ratio by isotopic methods has been questioned (Korkes, 1952), but if independent confirmation was found, then some mechanism of energy transfer must exist other than by the coupled reactions described on previous pages. Szent-Gyorgyi (1957) has recently discussed the possible transfer of excitation energy in biological systems. A well-known example is photosynthesis in which radiation produces electronic excitation in chlorophyll which is then converted into stabilised forms of energy.

Hill & Scarisbrick (1940) demonstrated that isolated chloroplasts evolve oxygen when illuminated in the presence of an artificial acceptor such as ferricyanide:

$$H_2O + A \xrightarrow{\text{light}} H_2A + \tfrac{1}{2}O_2.$$

The recent studies of Arnon, Whatley and Allen (1958) have shown that the Hill reaction is accompanied by phosphorylation. In the absence of carbon dioxide, illuminated chloroplast fragments, supplemented with an aqueous extract of chloroplasts, ADP, Pi and TPN, carry out the reaction

$$2ADP + 2Pi + 2TPN^+ + 4H_2O \rightarrow 2ATP + O_2 + 2TPNH + 2H^+.$$

Thus the excitation energy of chlorophyll can be used for the simultaneous reduction of TPN and phosphorylation of ADP. When FMN or vitamin K is added to the chloroplast fragments, the rate of photochemical phosphorylation is greatly increased, but the evolution of oxygen is abolished. This type of phosphorylation, termed cyclic phosphorylation, can be explained on the basis that the extra ATP is formed during the oxidation of TPNH by (OH), the oxidised product of the photodecomposition of water.

Arnon's experiments have demonstrated that light energy is utilised in the formation of ATP when carbon dioxide is low or absent. It is possible that these reactions take place during the midday closure of stomata, and photochemically derived ATP may be involved in many anabolic reactions. Thus, for example, Maclachlan and Porter (1959) have shown that the synthesis of starch from glucose (a reaction which requires ATP) takes place in illuminated leaves maintained under a low pressure of oxygen.

It will be noted that nothing has been said about the way in which the excitation energy of chlorophyll becomes stabilised. Szent-Gyorgyi suggests that excitational energy, denoted by E*, may be the core of reactions in which the living machinery is *driven* and work is produced, whereas the energy transferred by the group-transfer reactions of intermediary metabolism, denoted by E, is the core of reactions in which the living machinery is *built*. Mitochondria are the intracellular sites of intense energy transfer, and if excitation energy (E*) is involved in biological energy transfer, then it may be involved in oxidative phosphorylation. Szent-Gyorgyi has demonstrated that frozen solutions of riboflavin do not phosphoresce when irradiated with ultraviolet light, unless oxygen is present. He suggests that oxygen makes the triplet state of riboflavin unstable. Certain compounds quench the phosphorescence of riboflavin solutions and these compounds also separate the processes of oxidation and phosphorylation. For example, 2:4-dinitrophenol, sometimes used as a slimming agent, has been shown to uncouple oxidative phosphorylation at concentrations as low as 10^{-4} to 10^{-5}M, and at these concentrations it also quenches the phosphorescence of riboflavin.

The uncoupling properties of dinitrophenol have been demonstrated with a wide variety of tissues, but in terms of a group-transfer theory of oxidative phosphorylation no generally accepted explanation has been found. Cohn and Drysdale (1955) observed that the isotopic exchange between ^{18}O-labelled phosphate and water, which takes place during oxidative phosphorylation is inhibited by dinitrophenol, and they suggest that dinitrophenol exerts its effect before inorganic phosphate participates in oxidative phosphorylation.

Whatever the mechanism of action of dinitrophenol may be, its uncoupling properties have been extensively used in studies on the regulation of the rate of respiration. Experiments have established that at least two respiratory systems function in plant respiration, one of which is coupled to, and frequently limited by, phosphorylation.

Many experiments have been directed towards the demonstration that particular respiratory patterns are intimately connected with certain work processes. For example, Commoner and Thimann (1941) observed that at low concentrations, iodoacetate completely inhibits the growth of Avena sections while inhibiting respiration by less than 10 %. They concluded that a small fraction of the total respiration is different in kind and directly related to growth. Experiments showing growth responses to organic acids were interpreted on the basis that the fraction of respiration related to growth was the metabolism of the organic acids. Cooil (1952) demonstrated that under the conditions employed by Commoner and Thimann, the growth response observed was due to the use of potassium salts, since inorganic potassium salts produced a response and the sodium salts of the organic acids were without effect. However, Cooil observed that when the potassium requirement was met, high concentrations of organic acids could reverse the growth inhibition produced by iodoacetate, and so it appears probable that the metabolism of organic acids is related to the growth process. Dinitrophenol stimulates plant respiration (table 8), inhibits growth (Bonner, 1949), abolishes water uptake (Hackett and Thimann, 1953), and available evidence suggests that the common factor is the phosphorylating system associated with the oxidations of the Krebs cycle.

TABLE 8. *Stimulation of plant respiration by dinitrophenol*

Tissue	Respiration of tissue in presence of dinitrophenol as a percentage of control
Wheat roots	140
Tobacco leaves	165
Oat coleoptile	197
Potato	260

It has often been debated whether a work process such as salt uptake is the cause or the effect of increased respiration. The investigations on coupled oxidative phosphorylation show that we cannot separate cause and effect, the regulatory control of respiration being achieved because the performance of work involves dephosphorylation which leads to a compensatory stimulation of respiration and the synthesis of organic phosphates.

Perhaps the most thoroughly investigated example of respiratory rate control is the Pasteur effect, which may be defined as 'the action of oxygen in diminishing carbohydrate destruction' (Dixon, 1937), or as 'oxygen inhibition of fermentative processes' (Burk, 1939). The physiological significance of the Pasteur effect can be readily appreciated by reference to an analogy proposed by Lipmann (1942). An electric generating plant is designed to use a 'cheap' source of power such as water power, but to cover emergencies such as drought a more expensively operating steam-engine is built into the plant and a mechanism is provided to switch from one power supply to the other. In the cell, the cheap source of power is respiration—or more specifically oxidative phosphorylation—which produces 38 moles of ATP per mole of glucose, and the expensive power source is fermentation which produces only 2 moles of ATP per mole of glucose. The Pasteur effect is the switch mechanism between fermentation and respiration.

The Pasteur effect was observed in apples by Blackman and Parija (1928) and explained by Blackman as due to the existence of an anabolic reaction conserving carbon in air. This concept of oxidative anabolism has been widely accepted by plant physiologists, though there has been little direct evidence to support it, so that Turner (1951) stated in a review of the Pasteur effect that 'it is impossible at present to come to a definite conclusion regarding oxidative anabolism'. Recently, however, James and Slater (1959) have attempted to demonstrate oxidative anabolism with the aid of ^{14}C-labelled compounds, and these experiments led James (1957) to conclude 'that there is no evidence at present that a closed cycle of glycolysis, acid formation and carbohydrate resynthesis occurs on any massive scale in plant tissues in the absence of photosynthesis'.

An alternative explanation proposes that the Pasteur effect is due to a limitation of phosphate acceptor and/or inorganic phosphate with particular reference to the action of triosephosphate dehydrogenase:

$$\begin{array}{l} CH_2O\textcircled{P} \\ | \\ CHOH + ADP + Pi + DPN^+ \rightleftharpoons \\ | \\ CHO \end{array} \quad \begin{array}{l} CH_2O\textcircled{P} \\ | \\ CHOH + ATP + DPNH + H^+ \\ | \\ COOH \end{array}$$

The equation represents the activity of triosephosphate dehydrogenase and phosphoglycerokinase and shows that the rate of triosephosphate oxidation will depend on the concentration of ADP and Pi. Since respiration removes ADP and Pi it will tend to inhibit triosephosphate dehydrogenase and thereby inhibit fermentation. However, as pointed out by Lynen and Koenigsberger (1951), blocking triosephosphate dehydrogenase does not explain the decreased utilisation of carbohydrates observed in air, and the blocking would be expected to lead to an accumulation of fructose-1:6-diphosphate. Thus, additional factors must be involved, and it has been suggested (Lipmann, 1933) that the oxidation of an —SH group by oxygen might reversibly inhibit a glycolytic enzyme and a recent report (Turner and Mapson, 1958) suggests that glycolysis in cell-free extracts of peas is inhibited by oxygen. However, before any conclusions can be drawn on the mechanism of the Pasteur effect, information on the steady-state concentration of phosphorylated intermediates is required.

CHAPTER 4

Catabolism

THE catabolic reactions of carbohydrates, fats and proteins, can be considered to proceed in three phases (Krebs, 1953). In animals, the reactions of the first phase (fig. 25) take place to a large extent in the intestinal tract, but, with the exception of carnivorous plants, the reactions are intracellular in plants. The first phase produces a number of small molecules, which in the second phase are oxidised to carbon dioxide and a compound which is a reactant in the third phase—the Krebs cycle. An important feature pointed out by Krebs is that the common pathway of oxidation results in an economy of 'chemical tools'. Shortage of space limits discussion, but most of the biochemical principles can be illustrated by restricting the discussion to the reactions of carbohydrates and fats in Phase II and to the Krebs cycle.

CARBOHYDRATE METABOLISM

Prior to about 1934, there was considerable uncertainty concerning the main pathway of carbohydrate metabolism, and methylglyoxal (pyruvic aldehyde) was thought by many to be a precursor of alcohol and the immediate precursor of lactic acid. With regard to higher plants, Neuberg and his co-workers demonstrated the presence in leaves of the glyoxalase enzyme system which effects the conversion of methylglyoxal to lactic acid, and also that leaf preparations can catalyse the formation of methylglyoxal from hexose phosphate. However, in 1932 Lohmann demonstrated that a muscle preparation, freed of glutathione, but fortified with other cofactors, was unable to convert methylglyoxal to lactic acid, but was, nevertheless, able to form lactic acid from glycogen. This evidence, together with the observation that the glyoxalase system converts methylglyoxal to the 'unnatural' or $D(-)$ form of lactic acid (possibly in some cases to a racemic mixture), was sufficient to exclude methylglyoxal as a glycolytic intermediate. In passing it is

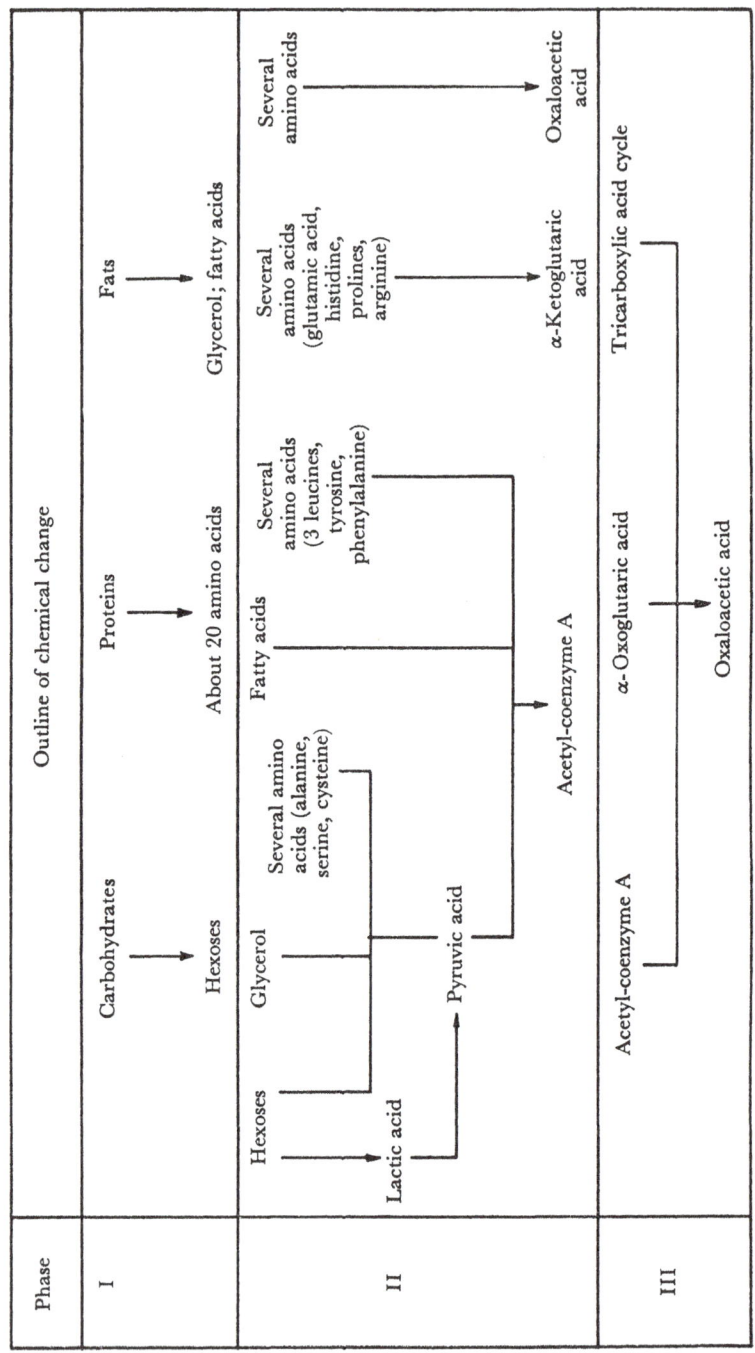

Fig. 25. The three phases of catabolism. After Krebs (1953).

54

interesting to note that it is now known that the 'unnatural' form of lactic acid is oxidised by animal mitochondria. In 1932 DL-glyceraldehyde-3-phosphate was synthesised by Fischer and Baer, and in 1933 the participation of 3-phosphoglyceraldehyde and 3-phosphoglyceric acid in glycolysis was recognised, so that in 1934 Embden & Jost were able to formulate a scheme for glycolysis essentially similar to the scheme shown in fig. 4. Evidence to support the Embden–Meyerhof pathway rapidly accumulated, and the scheme was thought to represent the main, if not the sole, pathway of carbohydrate metabolism. Evidence that the Embden–Meyerhof pathway functions in higher plants was provided by Tanko in 1936 and by James and his co-workers in 1941, but it was not until 1953 that the list of glycolytic enzymes isolated from plants was completed. The great success of the Embden–Meyerhof scheme drew attention away from alternative pathways of carbohydrate metabolism. However, the discovery in Warburg's laboratory of dehydrogenases active with glucose-6-phosphate and 6-phosphogluconate suggested the possibility of an alternative pathway. Dickens (1938) continued these investigations and established that D-ribose-5-phosphate was a highly active metabolite, but his investigations were unhappily interrupted by the war.

Since 1950, investigations in a number of laboratories have established the existence of an alternative pathway of carbohydrate metabolism, variously called the direct oxidation pathway, the Warburg–Lipmann–Dickens pathway, the hexose-monophosphate shunt and the pentose-phosphate cycle. Evidence that the pentose-phosphate pathway is present in higher plants is summarised below:

(1) All the enzymes required for the operation of the pathway have been demonstrated in higher plants.

(2) The postulated intermediates are known to be present in plants.

(3) When dephosphorylated, but radioactive intermediates are supplied to plant material, they serve as respiratory substrates and radioactivity appears in other intermediates of the pathway.

(4) The last piece of evidence which is based on the ratio

$$\frac{\text{per cent yield of }^{14}CO_2 \text{ from glucose-6-}^{14}C}{\text{per cent yield of }^{14}CO_2 \text{ from glucose-1-}^{14}C},$$

usually called the Bloom and Stetten ratio, requires detailed examination. Oxidation of carbohydrate via the Embden–Meyerhof pathway and the Krebs cycle, should give equal rates of $^{14}CO_2$ production from glucose-1-^{14}C or -6-^{14}C, due to the fact that triosephosphate isomerase equilibrates the dihydroxy-acetone phosphate and phosphoglyceraldehyde produced by the cleavage of fructose-1:6-diphosphate. Values greater than unity for the C_6/C_1 ratio have been reported (Gibbs and Beevers, 1955), and may be explained by the existence of a metabolic pool of dihydroxyacetonephosphate possibly derived from fat, which accompanied by low isomerase activity would lower the specific activity of dihydroxyacetone phosphate with respect to phosphoglyceraldehyde. Since C-3 of phospho-glyceraldehyde is derived from C-6 of glucose, a high C_6/C_1 ratio is to be expected. An alternative explanation follows from evidence suggesting that the biosynthesis of xylan and the pentoses of hemicellulose involves the loss of C-6 from hexose. Values of less than unity for the C_6/C_1 ratio suggest that some sugar is being oxidised via the pentosephosphate pathway which rapidly removes the C-1 of glucose in the reaction

$$\text{6-phosphogluconate} + TPN^+ \rightleftharpoons \text{ribulose-5-phosphate} + H^+ + TPNH + CO_2.$$

The Bloom and Stetten ratio has been used to assess the quantitative significance of the pentose-phosphate pathway. Relatively few quantitative studies have been performed with plants and quantitative conclusions should be treated with caution because of uncertainties inherent in the method.

As shown in chapter 1, it is possible to produce large changes in the relative velocities of simultaneous reactions by small increases in the concentration of the common substrate. The Embden–Meyerhof and pentose-phosphate pathways may be considered as simultaneous reactions taking place at the level of glucose-6-phosphate, so that the feeding of glucose to a tissue may increase the concentration of glucose-6-phosphate and possibly produce a large change in the relative velocities of the pathways.

The Bloom and Stetten procedure measures only the initial production of $^{14}CO_2$, because if the glucose were completely oxidised, the C_6/C_1 ratio would be unity. The necessity to measures initial rates prevents the formation of an isotopic steady state and makes it impossible to determine the velocity of a metabolic pathway by measuring the amount of isotope in the end-product. The problem is related to the passage of a solute through a reservoir of solvent. Suppose three men are positioned on the banks of a branched stream as shown in fig. 26. The man at position (1) adds a dye to the stream and the men at positions (2) and (3) measure the amount of dye passing them. It is clear that the pond will dilute the dye and so reduce the rate at which dye passes position (2). To calculate the rate of water-flow through the pond from the amount of dye passing

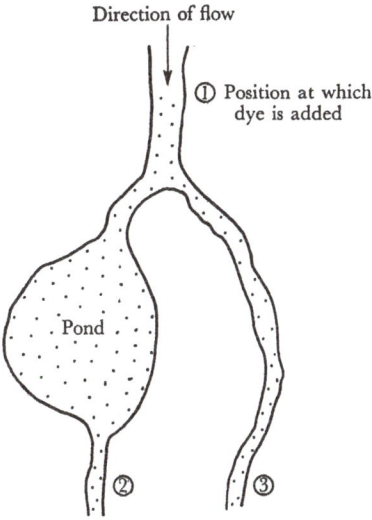

Fig. 26. Diagrammatical representation of the effect of pool size on the experimental determination of flow-rates.

position (2), it is necessary to know the volume of the pond and the conditions of mixing. If we assume instantaneous mixing, the problem is simplified and related to the mixing of solutions required for gradient elution in chromatography for which a general equation has been derived (Bock and Ling, 1954). For the case where the rate of flow into the pond equals the rate of flow out, the concentration of dye passing position (2) is given by

$$C = C_0\left(\frac{e^{v/V} - 1}{e^{v/V}}\right).$$

57

(C_0 is the concentration of dye entering the pond of volume V, C is the concentration of dye leaving the pond when the volume v has passed through the pond.) The volume of solution passing through the pond cannot be determined by measuring C_0 and C unless the volume of the pond is known. Similarly, on feeding a labelled compound to a plant, say glucose ^{14}C, a determination of the yield of radioactive carbon dioxide does not permit the estimation of the rate at which glucose is oxidised unless the concentrations of the intermediate between glucose and carbon dioxide are known. This effect could be of particular significance in the determination of the C_6/C_1 ratio which is weighted in favour of the pentose-phosphate pathway. The Embden–Meyerhof pathway equates C_1 and C_6 so that the C_6/C_1 ratio is, in effect, a comparison between the carbon dioxide produced from the C-1 of glucose by the two pathways. The pentose-phosphate pathway rather directly converts the C-1 of glucose to carbon dioxide, but the Embden–Meyerhof pathway converts the C-1 of glucose to the methyl group of pyruvate, and two complete turns of the Krebs cycle must be completed before the methyl group of pyruvate can be released as carbon dioxide. Consequently the Embden–Meyerhof pathway will tend to be underestimated. On the other hand, reactions which tend to equilibrate the C_6 and the C_1 atoms of glucose tend to underestimate the participation of a pentose-phosphate pathway.

Workers in the field of animal biochemistry have replaced the simple C_6/C_1 ratio with more complex formulations and recently the effect of recycling in the pentose-phosphate pathway has been examined (Wood and Katz, 1958). There is room to doubt the existence of cycling in the pentose-phosphate pathway (see p. 21), but it would be difficult to find exception to the statement of Wood and Katz that 'circumstances may arise which make the interpretation of tracer patterns very difficult both in terms of mechanisms or in terms of the quantitative role of the pentose-phosphate cycle in metabolism'.

(see p. 21)

FATTY ACID METABOLISM

Early attempts to detect intermediates in the metabolism of fatty acids were unsuccessful. However, in 1904 Knoop conceived the idea of introducing into the terminal methyl group of fatty acids a radical which was not easily changed in the body, so that the acid or its degradation products could be readily identified in the urine. He fed substituted phenyl derivatives of fatty acids containing 1–5 C-atoms to dogs, and examined the urine. Depending on the compound fed, one of two conjugates with glycine could be detected in the urine:

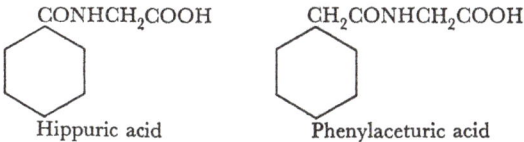

Hippuric acid was formed when a substituted fatty acid had an even number of methylene groups, and phenylaceturic acid when the number of methylene groups was added. Knoop, and independently Dakin, suggested that these results could be explained in terms of a process of β-oxidation, that is, by the successive removal of acetate from the fatty-acid chain. The mechanism proposed may be represented as below:

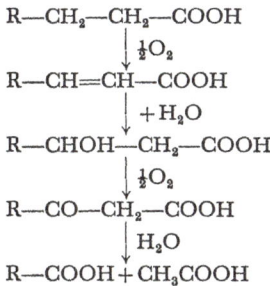

Recent work has shown that this scheme is essentially correct, though CoA derivatives rather than the free acids are involved.

Experiments with weed-killers, basically similar to the feeding experiments of Knoop, provided the first indication that the process of β-oxidation occurs in higher plants. Thus Grace (1939) demonstrated that the growth-promoting activity of a series of substituted aryl carboxylic acids showed no alternation of activity as the homologous series was ascended. Synerholm and Zimmerman later prepared a series of seven ω-2:4-dichlorophenoxyalkylcarboxylic acids and observed that only those acids with an odd number of methylene groups in the side chain produced epinastic responses in tomato plants. They suggested that the β-oxidation of compounds with an odd number of side-chain methylene groups would produce 2:4-dichlorophenoxyacetic acid, which is known to produce an epinastic response, whereas β-oxidation of acids containing an even number would yield the phenol without growth-regulating activity. Fawcett, Ingram and Wain (1954) supplied a series of ω-phenoxyalkylcarboxylic acids to flax plants and found that it was subsequently possible to isolate phenol only from plants treated with acids possessing an even number of methylene groups.

Although such evidence suggested the existence of a β-oxidation mechanism in plants, it was not until 1956 that Stumpf and Barber were able to demonstrate the oxidation of long-chain fatty acids in a cell-free system. Freshly prepared peanut mitochondria oxidised straight-chain fatty acids from acetate to stearate, provided the concentration of fatty acids was low and CoA, ATP, DPN, TPN, Mn^{2+} glutathione and a Krebs cycle acid were added as co-factors. These co-factor requirements are the same as those demonstrated as essential for the β-oxidation of fatty acids by a soluble system obtained from animal mitochondria (Green, 1954). The detailed reactions of β-oxidation in animals have been shown to be very similar to those postulated by Knoop and Dakin, the main modification being the activation of fatty acids so that the CoA derivatives rather than the free acids are involved in β-oxidation. The individual enzyme reactions are listed below:

(1) The formation of a fatty acyl-CoA derivative by the reaction

$$R—CH_2—CH_2—CH_2—COOH + ATP + CoA$$
$$\rightleftharpoons R—CH_2—CH_2—CH_2—COSCoA + AMP + PP.$$

Three fatty-acid activating enzymes are known, differing in their specificity with regard to optimum chain length, but over-lapping in their specificity, so that acids from C_2 to C_{16} can be activated.

(2) The oxidation of fatty acyl-CoA to α, β-unsaturated fatty acyl-CoA according to the reaction

$$R—CH_2—CH_2—CH_2—COSCoA + FAD$$
$$\rightleftharpoons R—CH_2—CH=CH—COSCoA + FADH_2.$$

Three fatty acyl-CoA dehydrogenases are known differing in their substrate specificity. All are metalloflavoproteins, though

60

the identification of copper as an essential component of the flavoprotein active with fatty acyl-CoA derivatives from C_4 to C_8 appears to be wrong (Beinert and Steyn-Parvé, 1957).

(3) The hydration of α, β-unsaturated fatty acyl-CoA to β-hydroxyacyl CoA

$$R—CH_2—CH{=}CH—COSCoA$$
$$\overset{H_2O}{\rightleftharpoons} R—CH_2—CH(OH)—CH_2—COSCoA.$$

The enzyme unsaturated acyl-CoA hydrase, also known as crotonase, acts upon all unsaturated acyl-CoA derivatives tested. Crotonase catalyses the hydration of both *cis*- and *trans*-crotonyl-CoA, the products being D($-$)β-hydroxybutyryl-CoA and L($+$)β-hydroxybutyryl-CoA respectively. The stereospecificity of crotonase with respect to a three-point attachment of the substrate to the enzyme has been discussed by Wakil (1956).

(4) The oxidation of β-hydroxyacyl-CoA to β-ketoacyl-CoA

$$R—CH_2—CHOH—CH_2—COSCoA + DPN^+$$
$$\rightleftharpoons R—CH_2—CO—CH_2—COSCoA + DPNH + H^+.$$

A single DPN specific enzyme β-hydroxyacyl-CoA dehydrogenase is active with all L($+$)hydroxyacyl-CoA derivatives tested. A D($-$)hydroxybutyryl-CoA dehydrogenase has also been demonstrated (Wakil, 1955).

(5) The cleavage of β-ketoacyl-CoA derivatives according to the reaction

$$R—CH_2—CO—CH_2—COSCoA + CoASH$$
$$\rightleftharpoons R—CH_2—COSCoA + CH_3COSCoA.$$

The enzyme β-ketoacyl-thiolase appears to be active regardless of chain length.

Though none of these enzymes has been purified from plant material, the activities of fatty-acid activating enzyme, butyryl-CoA dehydrogenase, crotonase and thiolase have been demonstrated in extracts of peanut mitochondria (Stumpf and Bradbeer, 1959). Additional evidence has been obtained to support the view that the process of fatty-acid oxidation catalysed by peanut mitochondria is similar to that found in animals. Thus malonate inhibits the oxidation of butyrate, presumably by inhibiting succinoxidase and so preventing the operation of the Krebs cycle. Further, the labelling of Krebs cycle acids

61

expected from the β-oxidation of butyrate-1-^{14}C has been demonstrated as well as the expected pattern of release of ^{14}C-labelled carbon dioxide from palmitic acid labelled in the second or third carbon atom. The β-oxidation of palmitate-3-^{14}C would be expected to give carboxyl-labelled acetyl-CoA, whereas palmitate-2-^{14}C would be expected to give methyl-labelled acetyl-CoA. Assuming that citrate formed from acetyl-CoA and oxaloacetate reacts asymmetrically with aconitase in the same way as in animals, carboxyl-labelled acetyl-CoA should give rise to labelled carbon dioxide after two turns of the cycle, but three turns would be necessary to release the label from methyl-labelled acetyl-CoA. Thus palmitic acid-3-^{14}C could give rise to labelled carbon dioxide before palmitic acid-2-^{14}C and the experimental confirmation (table 9) supports the concept of β-oxidation.

TABLE 9. *Pattern of labelling of ^{14}CO$_2$ produced from labelled fatty acid* (after Stumpf and Barber (1956))

Substrate	Percentage total ^{14}C in BaCO$_3$ after		
	30 min	60 min	120 min
Acetate-1-^{14}C	4·4	12·0	36
Acetate-2-^{14}C	2·2	6·3	22
Ratio odd/even	2·0	1·9	1·6
Palmitate-3-^{14}C	18·0	36·0	53
Palmitate-2-^{14}C	1·1	4·2	15
Ratio odd/even	16·5	8·6	3·5

In passing, it may be noted that at one time, citric acid was thought to be a side-product of the reactions of the Krebs cycle because labelling data indicated that the first formed tricarboxylic acid was, unlike citrate, an asymmetric substance. However, as Ogston (1948) pointed out, citrate could react asymmetrically if it formed a three-point attachment with the enzyme aconitase, and this suggestion was confirmed by Potter and Heidelberger (1949) who provided evidence that mammalian aconitase dehydrates citric acid, forming a double bond between the central carbon atom and the methylene carbon derived from oxaloacetate. The experiments have been widely accepted, despite the fact that the evidence is by no means clear-cut (see Heidelberger and Potter, 1951) and verification for the plant aconitase has not been attempted. The possibility that leaves may possess an aconitase dehydrating citrate between the central carbon atom and the methylene carbon derived from acetyl-CoA has been discussed by Nelson and Krotkov (1956). The experiments of Stumpf and Barber throw no light on this problem since carboxyl-labelled acetate will give rise to $^{14}CO_2$ before methyl-labelled acetate irrespective of the stereospecificity of the aconitase reaction.

Whilst attempting to demonstrate β-oxidation, Stumpf and his co-workers (see Stumpf and Bradbeer, 1959) detected an α-oxidation system in peanut cotyledons. The system is composed of a fatty-acid peroxidase which in conjunction with a system for generating hydrogen peroxide, peroxidatively decarboxylates fatty acids of chain length C_{14}—C_{18}, to an aldehyde and carbon dioxide:

$$Glycolate + O_2 \rightarrow glyoxalate + H_2O_2,$$

$$R—CH_2—CH_2—COOH + H_2O_2 \rightarrow RCH_2CHO + CO_2 + H_2O.$$

A DPN-specific aldehyde dehydrogenase, present in microsomes, oxidises the aldehyde to a carboxylic acid, which can then undergo a second cycle of α-oxidation:

$$R—CH_2CHO + DPN^+ \xrightarrow{H_2O} RCH_2COOH + DPNH + H^+.$$

The relative activities of the α- and β-oxidation systems is unknown, but since the peroxidase only attacks acids with chain lengths greater than C_{12}, the α-oxidation system may be regarded as a supplement to β-oxidation for the oxidation of long-chain fatty acids.

THE KREBS CYCLE

The metabolic cycle involved in the complete oxidation of pyruvate (fig. 8) and variously termed the Krebs cycle, the citric-acid cycle and the tricarboxylic-acid cycle, was proposed by Krebs in 1937. The view that the Krebs cycle is functional in plants, first proposed by Chibnall in 1939, is widely accepted. Evidence for the operation of the cycle in intact tissues has come from the careful analytical studies of potatoes at Cambridge (Barker and Mapson, 1955) and of tobacco leaves at Connecticut (Vickery and Palmer, 1957). However, following the demonstration that mitochondria isolated from mung bean seedlings oxidise the acids of the Krebs cycle (Millerd *et al.* 1951), many workers have directed their attention to the study of plant mitochondria (see Hackett, 1955). Though many workers were able to isolate mitochondria from seedlings which were capable of carrying out the complete oxidation of pyruvate, attempts to prepare active mitochondria from green leaves were at first unsuccessful (Brummond and Burris, 1954). More recently, active mitochondria have been isolated from a variety of green leaves (Smillie, 1955; James and Das, 1957). Considerable uncertainty remains concerning the quantitative significance of the Krebs cycle in the metabolism of leaves. Mesophyll cells contain relatively few mitochondria, and the distribution of radioactivity after feeding leaves with ^{14}C-labelled organic acids, whilst in agreement with the presence of a Krebs cycle, has been interpreted to indicate a slow rate of cycling (Zbinovsky and Burris, 1952). Calvin and his co-workers (see Bassham, Shibata, Steenberg, Bourdon and Calvin, 1956) have shown that in light little radioactivity from ^{14}CO$_2$ appears in tricarboxylic acids, but on changing to darkness radioactivity appears in citrate and glutamate. When once more the light is turned on, the label disappears from citrate and glutamate. These results have been interpreted to mean that light inhibits the reactions of the Krebs cycle. However, since Brown (1953) has shown by means of experiments with ^{18}O, that the respiration of *Chlorella* is unaffected by light, Bassham et al. (1956) conclude that the light inhibition of the transfer of 'photosynthetic' carbon to 'respiratory' carbon is restricted to reactions within the chloroplasts. Against this viewpoint is the finding that purified chloroplasts are unable to carry out the reactions of the Krebs cycle at a detectable rate (James and Das, 1957). A specific mechanism for the light inhibition, proposed by Calvin and Barltrop (1952), postulates that light maintains lipoic acid in the reduced state, thus preventing the participation of lipoic acid in the oxidative decarboxylation of α-oxoglutarate. However, the data of Bassham et al. (1956) indicates that light does not block the oxidation of α-oxoglutarate as would be expected on the basis of the explanation offered by Calvin and Barltrop. An alternative explanation suggested by Bidwell, Krotkov and Reed (1955) is that the lack of labelling in the Krebs cycle acids is due to lack of pyruvate rather than to the blocking of pyruvate oxidation. They argue that during photosynthesis the main movement of carbon is from phosphoglycerate towards sugars, and support their argument by demonstrating that photosynthetic carbon enters amino acids associated with the Krebs cycle, when the illuminated leaf is supplied with ammonium nitrate.

Despite the generally accepted importance of the Krebs cycle in plant metabolism, few attempts have been made to evaluate its quantitative significance. Millerd (1953) reported that mung bean hypocotyls respire at the rate of 150 mm³ O_2/h/g fresh weight, and mitochondria isolated from the same tissue oxidise succinate at the rate of 45 mm³ O_2/h/g original fresh weight. On the basis of an assumed 25 % recovery of mitochondria, Millerd concluded that mitochondria can oxidise succinate at a rate sufficient to account for the respiration of the hypocotyls. On the other hand, Price and Thimann (1954) have pointed out that since there are six dehydrogenases involved in the oxidation of carbohydrates, each dehydrogenase needs to proceed at only one-sixth of the total rate of oxidation. From their data they conclude that mitochondria contain more than adequate amounts of succinic and malic dehydrogenases to account for the respiration of pea internodes in terms of the Embden–Meyerhof pathway and the Krebs cycle.

The theoretical minimum rate at which mitochondria need oxidise succinate is, as Price and Thimann point out, one-sixth of the total, but by the same reasoning two substrates of the cycle—say succinate and malate—should be oxidised at one-third of the rate, whilst the complete oxidation of pyruvate by mitochondria should take place at five-sixths of the total rate. However, mitochondria do not oxidise two substrates at twice the rate of a single substrate (Davies, 1953), indeed, the oxygen uptake observed with one substrate is close to the maximum rate obtained with mixed substrates and is also close to the rate for the complete oxidation of pyruvate. The explanation appears to be that the rate of oxygen uptake is determined by the rate at which ATP formed by oxidative phosphorylation is dephosphorylated (Laties, 1953). It would thus appear that unless intact tissue possesses a more efficient phosphate-accepting system than the hexokinase 'trap' employed in studies of oxidative phosphorylation, mitochondria account for less than 50 % of the oxygen uptake of some plant tissues. It is, however, probable that mitochondria are damaged during isolation and full Krebs cycle activity may require enzymes not present in the mitochondria.

The failure to evaluate quantitatively the contribution to the total respiration of any metabolic pathway represents a major gap in the understanding of catabolism. This lack of quantitative information is reflected in the uncertainty associated with many physiological problems. For a proper understanding of cell physiology, quantitative information about the effect of various conditions on metabolic pathways is required, but many problems must be solved before biochemists can provide the required information.

CHAPTER 5

Anabolism

THE term anabolism is applied to those aspects of metabolism in which a process leads to the synthesis of more complex molecules. Synthetic reactions usually involve an increase in free energy and consequently synthetic reactions are half-reactions which must be coupled to a degradative reaction to provide the 'driving force' for synthesis. Consequently, anabolism requires the participation of catabolic reactions, and for this reason biochemists have until recently concentrated their efforts on the degradative aspects of metabolism. Chemists, on the other hand, have, as Robinson stated (1955), 'been tempted to leave the security of their own proper pastures and to graze, albeit speculatively, in the attractive fields of biochemistry'. These speculations are based on the detection of an architectural plan common to a group of natural products and on the supposition that the structural units comprising the natural products are related to the building units involved in biogenesis. This approach cannot define mechanistic details and the postulated building units must be considered as equivalent to and not identical with the actual building unit. Within these limits, considerable success has attended the speculations of organic chemists. For example, the recognition that the terpenes and camphors could be subdivided into C_5 units related to isoprene $CH_2{=}CH{-}C(CH_3){=}CH_2$ led to the formulation of two isoprene rules. Ruzicka and Stoll (1922) proposed that the biogenesis of terpenes results in molecules constructed of, or divisible into, isoprene units, and Robinson (1923) proposed that isoprene units are arranged in a 'head-to-tail' formation. These rules provide a unifying concept for compounds as different as the sterols, the carotenoids and rubber, and suggest a common biosynthetic pathway which has recently been supported by direct biochemical investigation on lycopene synthesis in tomatoes (Shneour and Zabin, 1959) and rubber synthesis in *Hevea brasiliensis* (Bandurski and Teas, 1957; Park and Bonner, 1958). Another example is the rapid advance towards an understanding of flavonoid biosynthesis (see Bogorad, 1958) which can in large measure be attributed to the theoretical contribution of Birch and Donovan (1953).

TABLE 10. *Examples of aldol condensations*

Compound with activated methylene group	Carbonyl compound	Product
Acetyl-CoA	Oxaloacetate	Citrate
Acetyl-CoA	Glyoxalate	Malate
Dihydroxyacetone phosphate	Phosphoglyceraldehyde	Fructose-1:6-diphosphate
Glycine	Hydroxymethyl THFA	Serine
Succinate	Glyoxalate	*Iso*citrate
Phosphoenolpyruvate	Erythrose-4-phosphate	2-keto-3-deoxy-D-arabo-heptonic acid-7-phosphate

Though the theoretical approach of the organic chemist has been invaluable, it must again be stressed that this approach cannot give information about mechanisms. Thus, because the laboratory synthesis of tropinone (Robinson, 1917) employs methylamine, it should not be taken to mean that methylamine is a postulated precursor of tropinone in the plant (cf. Cromwell, 1943). Similarly, we may recognise that many synthetic reactions found in plants involve an aldol-type condensation of a carbonyl group with an activated methylene group (table 10), but it would be unwise to suggest that identical enzymic mechanisms are involved.

The carbonyl group may be regarded as a 'hybrid' of the forms

$$C{=}O \qquad \overset{+}{C}{-}\overset{-}{O}$$

the positive carbonyl carbon being reactive towards nucleophilic reagents. For the formation of carbon to carbon bonds by aldol condensation, the second compound must contain a labile hydrogen atom activated by an electron-attracting group. In a base catalysed aldol condensation, the base facilitates the withdrawal of the proton, thereby creating a powerful nucleophilic reagent to attack the positive carbonyl carbon. The experiments of Rose and Rieder (1958) have shown that the enzyme catalysed aldol condensation of dihydroxyacetonephosphate and phosphoglyceraldehyde probably proceeds by an essentially similar mechanism. The enzyme aldolase, stereospecifically removes one of the two hydrogen atoms of the C-3

68

of dihydroxyacetonephosphate. The activation of this hydrogen precedes the condensation with phosphoglyceraldehyde, because aldolase catalyses the exchange of protons between dihydroxyacetonephosphate and the medium, in the absence of phosphoglyceraldehyde.

Certain aldol condensations do not involve this mechanism. Thus if the condensation of acetyl-CoA and oxaloacetate proceeds by a similar mechanism, condensing enzyme should be able to labilise one of the hydrogen atoms of the methyl group of acetyl-CoA and it should be possible to demonstrate an exchange reaction between the tautomers of acetyl-CoA:

$$CH_3-\overset{\overset{\displaystyle O}{\|}}{C}-SCoA \rightleftharpoons CH_2=\overset{\overset{\displaystyle O^-}{|}}{C}-SCoA + H^+$$

Experiments have shown that condensing enzyme does not catalyse this exchange and thus indicate a different mechanism for the aldol-type condensation.

The condensation of phosphopyruvate with erythrose-4-phosphate to form 2-keto-3-deoxy-D-arabo-heptonic acid-7-phosphate (an intermediate in the synthesis of the aromatic ring)

is catalysed by an enzyme which was first called 'phosphoenol-pyruvate-erythrose-4-phosphate aldolase' but because the condensation is thought to take place by a concerted mechanism, the name has been changed to '2-keto-3-deoxy-D-arabo-heptonic acid-7-phosphate-synthetase' (Srinivasan and Sprinson, 1959).

In recent years, much interest has been focused on the energetics of biosynthesis. An elementary treatment of thermodynamic principles has been given in chapter 3, and in this section these principles will be applied to the consideration of carbohydrate and fat synthesis.

One of the first enzymes to be discovered was amylase which hydrolyses starch to maltose:

$$\text{Starch} \underset{}{\overset{H_2O}{\rightleftharpoons}} n \text{ (maltose)}.$$

The equilibrium is so far to the right that the reaction may be considered irreversible, though the possibility that some synthesis occurs at interfaces has been considered by Oparin and Kurssanov (1931). The synthesis of starch possibly involves the reaction

$$n\text{-Glucose-1-phosphate} \rightleftharpoons \text{starch} + n\text{-}H_3PO_4,$$

which was first demonstrated by Hanes (1940). The equilibrium constant

$$K = \frac{(\text{Glucose-1-phosphate})^n}{(\text{Starch}) \, (\text{Pi})^n}$$

may be determined from the equilibrium ratio of glucose-1-phosphate to Pi, because the molar concentration of starch may be considered constant. At pH 7·0, equilibrium is attained when the ratio of glucose-1-phosphate to Pi is approximately 0·32, showing that the reaction is readily reversible and suggesting that possibly the breakdown of starch occurs via the enzyme phosphorylase, rather than by the action of amylase. It will be noted that the equilibrium ratio of glucose-1-phosphate to Pi varies with the pH, being 0·09 at pH 5·0, 0·15 at pH 6 and 0·32 at pH 7 (Hanes, 1940). Thus a high pH favours starch breakdown, as required by some theories of stomatal movement (see Williams, 1954). It has been argued (see Edelman, 1956) that amylase, which is highly active in plant extracts, does not exert its hydrolytic properties in the living cell, so that phosphorylase has a catabolic and an anabolic function. We note, in the slightly teleological terms, so dear to the heart of most biochemists, that the energy of the glucosidic linkage of starch, which is lost on hydrolysis to maltose, is preserved in the ester phosphate bond of glucose-1-phosphate. In this connection, it is of interest to note that doubt is now being cast upon the anabolic function of phosphorylase. The discovery by Leloir and Cardini (1957) of an almost irreversible synthesis of glycogen from uridinediphosphate glucose and the demonstration that phosphorylase activation invariably led to glycogen breakdown and never to glycogen synthesis (Rall and Sutherland, 1958), has led to the view that phosphorylase has a catabolic rather than an anabolic function. Glycogen synthesis from uridine-diphosphate glucose has been found to be associated with a particulate fraction obtained from muscle homogenates (Robbins, Traut and Lipmann, 1959). These authors also point out the thermodynamic advantages of glycogen synthesis by way of uridinediphosphate glucose, since the transfer of a glucose unit from uridinediphosphate glucose to glycogen should have a standard free-energy charge of -3 kcal, so that conversion to glycogen should be greater than 99 %. The participation of uridinediphosphate glucose in starch synthesis remains to be demonstrated.

The synthesis of glycogen from lactate in animal tissues was, until 1954, widely believed to take place by a reversal of the reactions of glycolysis (see Fruton and Simmonds, 1953). However, in 1954 Krebs calculated that an ATP/ADP ratio of the order 10^{16} was necessary to effect an appreciable synthesis of glycogen from lactate. Krebs, therefore, proposed that since three reactions of glycolysis are not readily reversed, at least three alternative reactions are involved in the synthesis of glucose from lactate. The three reactions cited by Krebs were:

(1) Glucose $+$ ATP $\rightleftharpoons$ glucose-6-phosphate $+$ ADP $+$ H$^+$,
$\Delta G' = -5 \cdot 1$ kcal.[1]
(2) Fructose-6-phosphate $+$ ATP
$\rightleftharpoons$ fructose-1:6-diphosphate $+$ ADP $+$ H$^+$, $\Delta G' = -4 \cdot 2$ kcal.
(3) Phosphoenolpyruvate $+$ ADP $+$ H$^+$ $\rightleftharpoons$ pyruvate $+$ ATP,
$\Delta G' = -6 \cdot 1$ kcal.

It will be noted that the high ATP/ADP ratio which favours the phosphorylation of pyruvate is unfavourable for the dephosphorylation of the sugar phosphates. Alternative reactions for (1) and (2) are:

(1 a) Glucose-6-phosphate $\xrightarrow{\text{H}_2\text{O}}$ glucose $+$ Pi, $\Delta G' = -3 \cdot 8$ kcal.
(2 a) Fructose-1:6-diphosphate $\xrightarrow{\text{H}_2\text{O}}$ fructose-6-phosphate $+$ Pi,
$\Delta G' = -4 \cdot 7$ kcal.

It is probably significant that specific phosphatases exist for the above reactions and specific phosphatases for other glycolytic intermediates are unknown.

[1] All values of $\Delta G'$ cited in this chapter are for standard conditions except that the standard condition of the H$^+$ ion is that of pH 7 instead of pH 0 (1 molal activity).

The reversal of reaction (3) involves the following reactions:

(3 a) Pyruvate + TPNH + CO_2 ⇌ malate + TPN^+,
         $\Delta G' = -0.3$ kcal.
(3 b) Malate + DPN^+ ⇌ oxaloacetate + DPNH + H^+,
         $\Delta G' = +7.1$ kcal.
(3 c) Oxaloacetate + ATP ⇌ phosphopyruvate + ADP + CO_2,
         $\Delta G' = -0.9$ kcal.
         Pyruvate + ATP + TPNH + DPN ⇌ phosphopyruvate
                + ADP + TPN + DPNH, $\Delta G' = +5.9$ kcal.

The free energy change for the overall reaction is only slightly less than for the direct phosphorylation of pyruvate by ATP ($\Delta G' = +6.1$ kcal). The difference of 0.2 kcal is due to the fact that the redox potential of TPN at pH 7 is 4 mV more negative than that of DPN so that the reaction

$$TPNH + DPN \rightleftharpoons TPN + DPNH$$

has a free energy change of -0.2 kcal. However, the overall reaction can be 'driven' towards phosphopyruvate, because the oxidation of DPNH is a highly exergonic reaction ($\Delta G' = -52.4$ kcal), which means that in the presence of oxygen and the appropriate enzymes, the ratio DPN/DPNH will be high and so drive the reactions in the direction of phosphopyruvate synthesis. Similarly, if other reactions in the cell can maintain a high TPNH/TPN ratio, phosphopyruvate synthesis will be favoured.

According to this scheme, carbon dioxide fixation is essential for the reversal of glycolysis and the observed labelling of liver glycogen by radioactive carbon dioxide can be readily explained if malate is in equilibrium with fumarate. Similarly the scheme explains the randomisation of radiocarbon found in liver glycogen formed from ^{14}C-labelled pyruvate. By the same token, the observation that ^{14}C-labelled pyruvate is incorporated into glycogen by diaphragm muscle without rearrangement of the carbon skeleton (Hiatt, Goldstein, Lareau and Horecker, 1958), argues against the operation of the scheme in this tissue. The demonstration of a net synthesis of phosphopyruvate from pyruvate and ATP in a system consisting of pyruvate kinase and mitochondria (Krimsky, 1959), suggests the possibility of a limited synthesis of glycogen by the direct reversal of glycolysis. In this connection, it should be noted that the ATP/ADP ratio of 10^{16} which Krebs calculated was necessary for a net synthesis of glycogen, is an overestimate. The ratio ATP/ADP necessary for the production of equimolar concentrations of glycogen and lactate can be calculated as follows:

$$\text{Glycogen} + H_2O \rightleftharpoons 2\,\text{lactate}^- + 2H^+, \quad \Delta G^0 = -32 \cdot 2 \text{ kcal,}$$

$$3ADP^{3-} + 3HPO_4^{2-} + 3H^+ \rightleftharpoons 3ATP^{4-} + 3H_2O,$$
$$\Delta G^0 = -2 \cdot 7 \text{ kcal,}$$

$$\text{Glycogen} + 3ADP^{3-} + 3HPO_4^{2-} + H^+$$
$$\rightleftharpoons 2\,\text{lactate} + 3ATP^{4-} + 2H_2O, \quad \Delta G^0 = -34 \cdot 9 \text{ kcal,}$$

$$\Delta G^0 = -RT\ln K, \quad 34 \cdot 9 = 1 \cdot 36 \log K, \quad K = 10^{25 \cdot 7}.$$

At pH 7 and with all reactants except ATP + ADP at 1 molal activity

$$K = \frac{(\text{lactate})^2 (ATP)^3}{(\text{glycogen})(ADP)^3 (Pi)^3 H^+} = 10^{25 \cdot 7},$$

$$\frac{(ATP)^3}{(ADP)^3 \times 10^{-7}} = 10^{25 \cdot 7},$$

$$\frac{ATP}{ADP} = 10^{6 \cdot 2}.$$

When all reactants except ADP are at a concentration of 10^{-3} M

$$K = \frac{(10^{-3})^2 (10^{-3})^3}{10^{-3} (ADP)^3 (10^{-3})^3 (10^{-7})} = 10^{25 \cdot 7},$$

$$(ADP)^3 = 10^{-21 \cdot 7},$$

$$ADP = 10^{-7 \cdot 2},$$

$$\frac{ATP}{ADP} = 10^{4 \cdot 2}.$$

Whilst this ratio is still very high, it is not impossibly high, and suggests the possibility of a limited reversal of glycolysis, provided mitochondria can maintain an ATP/ADP ratio of about 10^4. That this ratio is within the capacity of mitochondria is suggested by the results of Von Korff and Twedt (1957) who found that under aerobic conditions, mitochondria can maintain the ratio of phosphopyruvate to pyruvate at about 10.

The equilibrium constant for the pyruvate kinase reaction is

$$\frac{(ATP)(\text{pyruvate})}{(ADP)(\text{phosphopyruvate})} = 4 \cdot 5 \times 10^3,$$

so that the calculated ratio of ATP/ADP would be $4 \cdot 5 \times 10^4$, which is adequate for a limited reversal of glycolysis.

73

Turning now to the related synthesis of starch in plants, we find that much attention has been given to this subject in relation to the Pasteur effect. The explanation of the Pasteur effect offered by Blackman requires the existence of an anabolic reaction for the resynthesis of carbohydrates. This explanation has been widely accepted by plant physiologists (see, for example, Bennet-Clark and Bexon, 1943), but tracer experiments by James (p. 51) suggest that in general carbohydrates are not formed from organic acids in the dark.

In special cases, however, carbohydrates are formed from organic acids and a good example is the incorporation of acetate

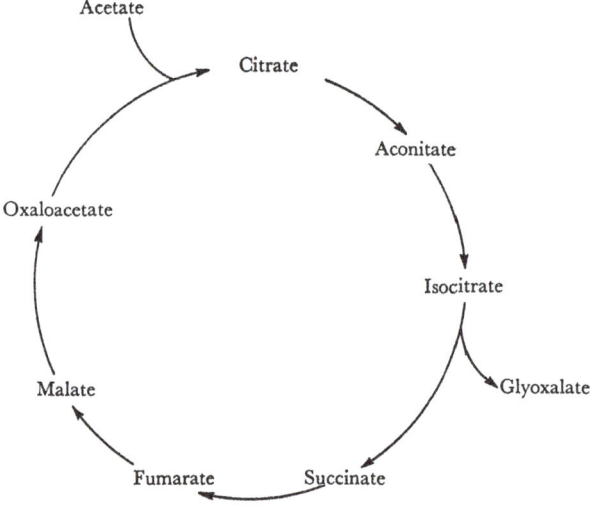

Fig. 27. Reactions of the glyoxalate cycle. The glyoxalate cycle, like the Krebs cycle, does not produce a net increase in cycle intermediates. However, a net gain of intermediates may be achieved by the aldol condensation of acetyl-CoA and glyoxalate catalysed by malate synthetase.

into sucrose by slices of castor beans (*Ricinus* sp.) (Beevers, 1957). The formation of carbohydrates from fats has long been known to occur in oil-containing seeds, but the mechanism of the reaction has only recently been proposed (Kornberg and Krebs, 1957) and supported by experiment (Kornberg and Beevers, 1957). Kornberg and Krebs have proposed a 'glyoxalate cycle' (fig. 27) which effects the oxidation of acetyl-CoA to glyoxalate. The glyoxalate formed by the cycle may then condense with acetyl-CoA to form malate which can give rise to carbohydrate by the reactions previously discussed. The two key enzymes, *iso*citritase and malate synthetase, have been demonstrated in castor beans by Kornberg and Beevers. The presence of an enzyme in plants forming phosphopyruvate from oxaloacetate and ATP has been demonstrated by Tchen and Vennesland (1955) and the formation of phosphopyruvate from malate has been demonstrated with mitochondria from pea seedlings (Davies, 1956). The presence of glycolytic enzymes is well documented and fructose-1:6-diphosphatase has been demonstrated in wheat (Edelman, 1958), and has been partially purified from spinach (Racker and Schroeder, 1958). It would thus appear that at least some plants have the necessary enzymes for the dark formation of starch from organic acids, and such a synthesis has been demonstrated in leaves of sunflower (Gibbs, 1951).

Fatty-acid synthesis has been considered as a reversal of β-oxidation, and the question of what determines the shortening of fatty-acid chains has been discussed by Krebs and Kornberg (1957). If we represent the overall process by the equation

$$C_n \text{Acyl CoA} + \text{DPNH}_2 + \text{reduced flavoprotein} + \text{acetyl CoA}$$
$$\rightleftharpoons C_{n+2} \text{Acyl CoA} + \text{DPN} + \text{flavoprotein} + \text{CoA}$$

the reaction will proceed in the direction of chain lengthening when the values for the three ratios

$$\frac{\text{Acetyl CoA}}{\text{CoASH}}, \quad \frac{\text{reduced flavoprotein}}{\text{oxidised flavoprotein}}, \quad \frac{\text{DPNH}_2}{\text{DPN}}$$

are relatively high.

The enzymes involved in β-oxidation are located in the mitochondria and since the ratios $\text{DPNH}_2:\text{DPN}$ and reduced:oxidised flavoprotein are probably low in mitochondria, it is unlikely that the reversal of β-oxidation takes place in mitochondria. Fat synthesis has, however, been demonstrated in the supernatant fraction from Avocado fruit, where TPNH, but not DPNH, was found to be necessary (Stumpf and Barber, 1957). Later work established that carbon dioxide was necessary for fat synthesis, but was not incorporated into fat. A probable explanation has been advanced by Brady (1958) who demonstrated the synthesis of long-chain fatty acids from acetaldehyde and malonyl-CoA in the presence of TPNH and Mn; —ATP not being required. In fatty-acid synthesis from acetyl-CoA, the carbon dioxide requirement probably relates to the carboxylation of acetyl-CoA to malonyl-CoA, a reaction which may involve 'activated' carbon dioxide. The non-incorporation of carbon dioxide into fatty acids follows from the decarboxylation which malonyl-CoA undergoes during the condensation with acetaldehyde:

$$
\begin{array}{l}
{}^{14}CO_2 \\
\qquad \searrow \\
CH_3COSCoA
\end{array}
\qquad
\begin{array}{l}
{}^{14}COOH \\
| \\
CH_2 \\
| \\
COSCoA
\end{array}
$$

$$R\text{---}CHO$$

$$R\text{---}CHOH\text{---}CH_2\text{---}COSCoA + {}^{14}CO_2$$
$$\downarrow -H_2O$$
$$R\text{---}CH\text{==}CH\text{---}COSCoA$$
$$\downarrow +[H]$$
$$R\text{---}CH_2\text{---}CH_2\text{---}COSCoA$$
$$\downarrow +TPNH$$
$$R\text{---}CH_2\text{---}CH_2\text{---}CHO + CoASH$$

These reactions are still tentative but the lengthening of the fatty-acid chain could occur by the successive condensation of aliphatic aldehydes with malonyl-CoA. The product of condensation would be a β-hydroxy derivative of CoA which could be dehydrated by crotonase or a similar enzyme. The reduction to fatty acyl-CoA probably involves flavoproteins. The reduction of the carboxyl group to the level of an aldehyde is possible provided the carboxyl group is activated and a TPN-specific dehydrogenase is probably involved.

From these specific examples we note that, in general, synthesis is not the simple reversal of degradation and that anhydrides play a part in most synthetic reactions. In general, the synthesis of the various anhydrides requires ATP, but in a number of cases alternative reactions are known. For example, acetyl-CoA may be formed according to the equation

$$\text{Acetate} + \text{ATP} + \text{CoASH} \rightleftharpoons \text{Acetyl-CoA} + \text{PP},$$

but may also be formed by the reaction

$$\text{Acetaldehyde} + \text{CoASH} + \text{TPN}^+ \rightleftharpoons \text{Acetyl-CoA} + \text{TPNH} + \text{H}^+.$$

Similarly formyl-THFA may be formed by the reaction

Formate + THFA + ATP ⇌ formyl-THFA + ADP + Pi,

or by the reaction

Hydroxymethyl-THFA + TPN$^+$
$$\rightleftharpoons \text{formyl-THFA} + \text{TPNH} + \text{H}^+.$$

Of the various anhydrides participating in biosynthesis, the nucleoside pyrophosphate group of compounds is particularly important. The first members of this group to be discovered were uridinediphosphate glucose and uridinediphosphate galactose which were shown to be intermediates in the interconversion of α-D-glucose-1-phosphate and α-D-galactose-1-phosphate (Caputto, Leloir, Trucco, Cardini and Paladini, 1949).

Uridinediphosphate glucose is formed by the reaction

glucose-1-phosphate + UTP ⇌ UDP glucose + PP

which is catalysed by the enzyme uridinediphosphate glucose pyrophosphorylase. The product of this reaction is involved in the biosynthesis of sucrose according to the reactions

UDP glucose + D-fructose-6-phosphate
$$\rightleftharpoons \text{sucrosephosphate} + \text{UDP}$$
and UDP glucose + D-fructose ⇌ sucrose + UDP.

Both reactions have been demonstrated in plants (Leloir, Cardini and Chiriboga, 1955) though in sugar-beet leaves the reaction leading to sucrose phosphate predominates, whereas in peas the glycosyl transfer from uridinediphosphate glucose can take place to fructose or its phosphate ester.

Uridinediphosphate glucose is also the probable precursor of arabans, xylans and pectin (fig. 28). The oxidation of uridinediphosphate glucose to uridinediphosphate glucuronic acid has been demonstrated with preparations from pea seedlings and the enzyme purified 1000-fold (Strominger and Mapson, 1957). The reaction is a two-step oxidation, but the postulated aldehyde intermediate has not been detected. The decarboxylation of uridinediphosphate glucuronic acid to yield uridinediphosphate xylose which was proposed by Altermatt and Neish (1956) has not been directly demonstrated, but a Walden inversion between uridinediphosphate xylose and uridinediphosphate arabinose has been demonstrated with plant extracts (Ginsberg, Stumpf and Hassid, 1956). Uridinediphosphate glucuronic acid and uridinediphosphate galacturonic acid are interconvertible by a Waldenase enzyme found in mung bean seedlings (Neufeld, Fengold and Hassid, 1958). The decarboxylation of UDP galacturonic acid has not been demonstrated, but is in accord with the labelling of the arabi-

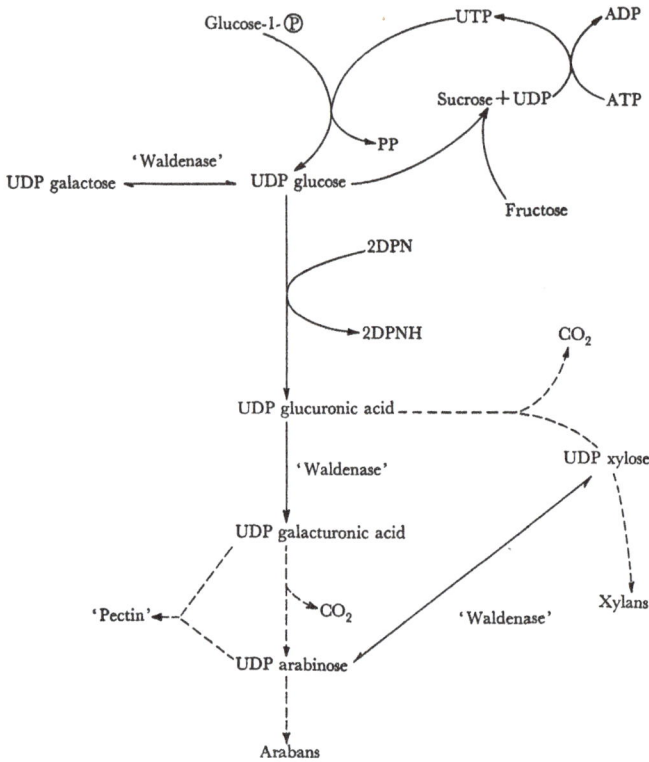

Fig. 28. Metabolism of uridinediphosphate carbohydrates. Evidence for reactions shown by dotted lines has been obtained in tracer experiments, but evidence with cell-free preparations is lacking. The various epimerisations shown in the scheme involve specific 'Waldenases'.

nose moieties of pectin observed by feeding labelled glucose to boysenberries (Seegmiller, Axelrod and McGready, 1955).

Space does not permit adequate discussion of this rapidly expanding field, or of the related cytidine diphosphate substrate compounds which are key reactants in the metabolism of phospholipids. Fortunately an excellent review by Baddeley and Buchanan (1958) is available.

To perform biosynthesis, the cell must provide anhydrous

high-energy compounds in an aqueous medium and highly reduced compounds in an oxidising system. The cell is able to provide both conditions because it possesses two pathways for oxidising carbohydrates, two types of pyridine nucleotide co-enzymes and membranes to separate the two systems. One pathway—glycolysis and the Krebs cycle—functions with DPN to catalyse the overall reaction

$$C_6H_{12}O_6 + 10DPN^+ + 2FAD^+ + 6H_2O$$
$$\rightleftharpoons 6CO_2 + 10DPNH + 12H^+ + 2FADH.$$

(FAD represents the flavin prosthetic group of succinic dehydrogenase.)

Of the 10 moles of DPN reduced per mole of glucose, 8 moles are reduced by reactions of the Krebs cycle occurring within the mitochondria. The double membranes of the mitochondria contain the electron transporting system and the associated system for oxidative phosphorylation. Consequently it is in the lipid-rich membranes that ATP—the most important high-energy anhydrous compound—is formed during the passage of electrons from DPNH and FADH to oxygen.

Situated outside the mitochondria, and thereby separated from the oxidising system, are the enzymes of the pentose-phosphate pathway. The two dehydrogenases of the pathway are specific for TPN and the pathway catalyses the overall reaction

$$C_6H_{12}O_6 + 12TPN^+ + 6H_2O \rightleftharpoons 6CO_2 + 12TPNH + 12H^+.$$

The reducing potential (TPNH) produced by the pathway, can be used to couple the oxidation of carbohydrates to a variety of reductive synthetic reactions.

Considered from a different point of view, the dual requirement for anhydrides and reducing potential provides an explanation for the existence of DPN and TPN specific enzymes. Thus the fixation of carbon dioxide by TPN specific malic enzyme is essentially the reverse of the DPN-linked oxidation of malate which occurs as part of the Krebs cycle. The simultaneous functioning of both systems is possible because different coenzymes are involved and because the systems are spatially separated.

Duality of pathways is a feature of the organisation of anabolic reactions. In cases where a catabolic reaction involves a large change in free energy, it is clear that a net synthesis cannot be achieved by a reversal of the catabolic reaction, and dual pathways must be present for catabolism and anabolism. It is now also becoming clear that in many cases where energetically it is possible for a single set of enzymes to function in catabolism and anabolism, dual pathways are nevertheless present. Thus the reversibility of the fatty-acid oxidation system discussed by Krebs and Kornberg (1957) is probably a theoretical rather than a practical consideration. The enzymes of the β-oxidation system probably function only in the direction of catabolism, synthesis being achieved by a different pathway involving the aldol condensation of an aldehyde with malonyl-CoA (see p. 76). Similarly, though the phosphorylase reaction is reversible it seems possible that there is a separate anabolic pathway involving uridinediphosphate glucose (p. 70).

These concepts lead to the speculation that DPN-specific dehydrogenases function in the direction of substrate oxidation, whereas with the exception of the dehydrogenases of the pentose-phosphate pathway, TPN-specific dehydrogenases function in the direction of substrate reduction. This speculation is no doubt an over-generalisation, and the most important exception is glutamic dehydrogenase which catalyses the reaction

$$\alpha\text{-Oxoglutarate} + \text{DPNH} + \text{H}^+ + \text{NH}_3$$
$$\rightleftharpoons \text{glutamate} + \text{DPN}^+ + \text{H}_2\text{O}$$

which is the only known reaction in plants, with the possible exception of the aspartase reaction, for producing amino groups. The reduction of nitrate to ammonia is catalysed by TPN-specific enzymes and the above generalisation requires the presence of a TPN-specific glutamic dehydrogenase to complete the pathway from nitrate to amino groups. Thus far only DPN-specific glutamic dehydrogenase has been detected in plants, but the existence of a TPN-specific triosephosphate dehydrogenase in plants has only recently been demonstrated and a TPN-specific glutamic dehydrogenase may await detection. In this connection, the observation of Kretovitch (1958), that the formation of glutamate in pea homogenates is stimulated by ATP, could be explained if ATP phosphorylates glucose to yield glucose-6-phosphate and so provide reduced TPN for a TPN-specific glutamic dehydrogenase. If the above views are correct it is of interest that 1 mole of ATP could bring about the fixation of 24 moles of ammonia. An alternative explanation would be that 'active' ammonia is formed from ATP and ammonia and evidence for this has been obtained in liver.

A further biological advantage following from the possession of DPN and TPN-specific enzymes, lies in what may be called the *push-pull* method of synthesis. Consider the reaction

$$UDPGal \rightleftharpoons UDPG.$$

The enzyme galactowaldenase, sometimes called 4-epimerase, requires DPN for the formal inversion of the 4-position in the hexose and the reaction mechanism is probably

$$\underset{H}{\overset{OH}{\diagup}} \underset{DPNH}{\overset{DPN}{\rightleftharpoons}} \underset{}{\overset{O}{\diagdown}} \underset{DPN}{\overset{DPNH}{\rightleftharpoons}} \underset{OH}{\overset{H}{\diagdown}}$$

The reaction involves an internal oxido-reduction, but because a single pyridine nucleotide is involved, the coenzyme cannot affect the equilibrium which is 3:1 in favour of the glucose nucleoside.

The formation of oxaloacetate from pyruvate and CO_2, involves an oxido-reduction, but because two pyridine nucleotides are involved, a *push-pull* effect can occur. This particular reaction is discussed on p. 72, but can be represented by the general equations

$$S + TPNH \rightleftharpoons SH + TPN,$$
$$SH:DPN \rightleftharpoons S^1 + DPNH.$$
$$Sum \quad S + DPN + TPNH \rightleftharpoons S^1 + DPNH + TPN.$$

The overall reaction $S \rightleftharpoons S^1$ is neither an oxidation nor a reduction, but the above mechanism allows the reaction to be '*pushed*' by maintaining DPN and TPNH in 'high' concentration and '*pulled*' by removing DPNH and TPN. Examples of this type of mechanism are shown in table 11.

TABLE 11. *Examples of 'Push-Pull' mechanism*

S + H⁺ TPNH SH₂ + DPN		SH₂ + TPN⁺ DPNA + H + S¹
S	SH$_2$	S^1
Pyruvate + CO_2	Malate	Oxaloacetate
L-Xylulose	Xylitol	D-Xylulose
Glucose	Sorbitol	Fructose
Glucuronate	L-Gulonate	3-Keto-L-gulonate

6   81   DP

CHAPTER 6

Links between Metabolic Pathways

IN previous chapters it has sometimes been necessary to discuss metabolic pathways without commenting on their interaction with other pathways. However, as Chibnall (1939) has pointed out 'Proteins, carbohydrates and organic acids are all three intimately connected with the living process of plants, and the metabolism of one of them cannot be satisfactorily investigated if concomitant changes in either or both of the other two are ignored'.

As an example of interaction between metabolic pathways we shall discuss the relationship between protein metabolism and respiration. Gregory and Sen (1939) considered that the amino acids arising from protein degradation in barley leaves suffered oxidative deamination, and they postulated that the rate of carbon dioxide production is controlled by the operation of a protein cycle. The concept of a protein cycle was greatly strengthened by work with ^{15}N which suggested that proteins were continually breaking down and being resynthesised, as, for example, was demonstrated for sunflower leaves by Hevesey, Linderstrom-Lang, Keston and Carsten (1940). However, investigations on the synthesis of adaptive enzymes in bacteria have suggested that protein turnover does not take place during the exponential phase of growth. Thus, for example, Rotman and Spiegelman (1954) have shown that the adaptive enzyme β-galactosidase is synthesised almost exclusively from the material that is assimilated after the addition of the inducer, and hence proteins existing in the non-induced bacteria are not precursors of the new protein. The existence of protein turnover in animals at the cellular level has also been questioned (Cohn, 1957). Recently the concept of a protein cycle has been expanded by Steward, Bidwell and Yemm (1958) to include the following ideas.

(1) The cell is heterogeneous, one phase being associated with assimilative metabolism and protein synthesis, the other with protein breakdown, catabolism and storage.

82

(2) A large part of carbon dioxide production arises from the oxidation of breakdown products arising from protein turnover.

(3) Free amino acids are not immediate precursors of protein.

The last proposition constitutes what has come to be known as the alternative path hypothesis of protein synthesis. The original suggestion of Abderhalden—the diketopiperazine hypothesis is no longer tenable and supporters of the alternative pathway have not specified any compounds which would function as the immediate precursors of protein. This proposition has been criticised by Yemm and Folkes (1958) and by the author (1959) and the evidence discussed on p. 30 strongly supports the Hofmeister–Fischer hypothesis of the step-by-step condensation of amino acids to form peptides, polypeptides and proteins.

The second postulate is essentially the hypothesis of a protein cycle which has recently been discussed by Folkes and Yemm (1958) who concluded that the respiratory behaviour of seedlings was consistent with the operation of a protein cycle but was equally consistent with the view that carbon dioxide was formed chiefly from the products of glycolysis. James (1957) concluded 'that there is no reason to suppose that the bulk of carbon emitted as carbon dioxide in plant respiration has passed through such a (protein) cycle and a good deal against it'. Steward and Pollard (1957), on the other hand, claim that the experiments of Vittorio, Krotkov and Reed (1955) show that though the addition of glucose-^{14}C to wheat leaves increases the respiration rate, the additional carbon dioxide is mainly unlabelled and they suggest that the extra carbon dioxide comes from the catabolic phase of a protein cycle, accelerated by the added sugar. However, an examination of the original data shows that approximately 80 % of the extra carbon dioxide produced is derived from glucose.

Despite the uncertainties of the physiological relationship between respiration and protein synthesis, biochemical interaction between protein and carbohydrate metabolism is assured because a number of the intermediates in carbohydrate metabolism are also intermediates in the metabolism of certain amino acids. Thus 3-phosphoglycerate, which is a well-known intermediate in the glycolytic pathway, has been shown by

Hanford and Davies (1958) to serve as the precursor of phosphoserine

3-phosphoglycerate + DPN$^+$
$\rightleftharpoons$ 3-phosphohydroxypyruvate + DPNH + H$^+$,
3-phosphoserine + α-oxoglutarate
$\rightleftharpoons$ 3-phosphohydroxypyruvate + glutamate.

Similarly, pyruvate, oxaloacetate and α-oxoglutarate are intermediates of glycolysis and the Krebs cycle, and are the precursors of the corresponding amino acids, alanine, aspartate and glutamate.

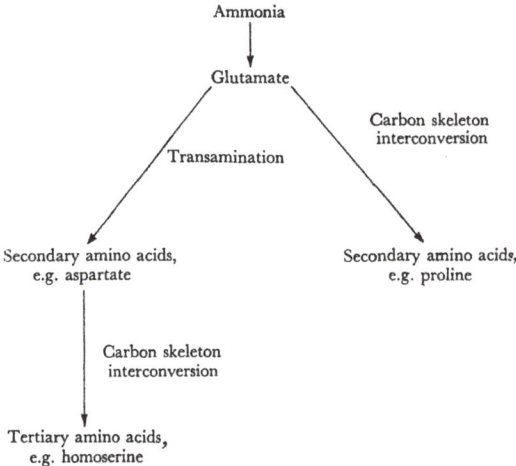

Fig. 29. Classification of amino acids in relation to the origin of the amino group. After Folkes (1959).

Considered from the viewpoint of the origin of the amino group, amino acids may be classified as shown in fig. 29. Kinetic evidence to support the view that glutamate is the most important primary product of ammonia assimilation in the yeast *Torulopsis utilis* has been presented by Folkes (1959). During the steady-state growth of a yeast population, Folkes has pointed out that the amount of the *primary* amino acid will increase exponentially

$$B = B_0 e^{at}$$

(where B is the quantity of the primary amino acid at time t and a is the growth-rate constant). If ammonia (A), the immediate precursor of B is labelled with ^{15}N so that it contains A^* atom

84

per cent excess ^{15}N, Folkes has shown that the atom per cent excess of ^{15}N in B is given by the expression

$$\frac{A*}{(A*-B*)} = e^{bt}$$

(where b is a coefficient involving the transfer coefficient for synthesis and utilisation of B and also the growth constant a). Thus a plot of

$$\log_e \frac{A*}{(A*-B*)}$$

against t should give a straight line of slope b passing through the origin. Data plotted in this way indicate that glutamate is a primary product of ammonia assimilation, but the presence of a lag indicates a secondary origin for all other α-amino acids with the exception of glutamine. Since glutamine is probably formed by the amidation of glutamate, it would be expected to have a secondary origin. However, as pointed out by Folkes the labelling of glutamine can be explained by assuming that glutamine is formed from glutamate at a site removed from the main glutamate pool.

It seems desirable at this point to note some of the difficulties involved in the analysis of kinetic experiments. The interpretation of kinetic data requires the formulation of a model or reaction mechanism from which a rate law can be derived and tested for closeness of fit to the experimental results. It follows that if the rate law is derived for a particular set of conditions, say the steady state, then the kinetic measurements must be made during the steady state. In recent years, steady-state enzyme kinetics have been applied to physiological problems such as salt uptake and the effects of auxin on cell elongation. In some cases the methodology has been inadequate, but even when the methods are adequate, there is frequently a tendency to overstate the conclusions of the kinetic analysis.

The application of steady-state kinetics to physiological problems is based on the finding that the kinetics of salt uptake and auxin-induced cell elongation bear a formal relationship to the Michaelis–Menten curve for the effect of substrate concentration on the velocity of an enzyme catalysed reaction. There is a natural tendency to force data into agreement with a preconceived hypothesis and it seems to the author that kinetic data are often used in this way. There is a large amount of evidence (see Bonner and Bandurski, 1952) suggesting that in order to act as a plant growth hormone, a molecule must possess a carboxyl group located in a side chain attached to an unsaturated cyclic nucleus and also a substitutable group of critical reactivity in the nucleus, *ortho* to the side chain. Consequently it is perhaps not surprising that kinetic experiments showing an auxin-induced inhibition of growth were interpreted by Foster, McRae and Bonner (1952) in terms of a two-point attachment of the auxin to a receptive entity in the cell. However, if we examine the derivation of the rate law given by these workers, we find that the same equation is obtained irrespective of the number of points of attachment between auxin and receptor. The basic assumption of the kinetic treatment given by Foster, McRae and Bonner is that the receptor molecule can combine with 1 molecule of auxin to give an active complex and also with 2 molecules to give an inactive complex. This assumption introduces the square of the auxin concentration into the denominator of the rate equation and thereby accounts for the observed optimum auxin concentration. The assumption of a two-point attachment is additional and unnecessary. The combination of a second mole of substrate at a neighbouring site may lead to inhibition or activation depending upon the effect upon the catalytic site. Following the nomenclature of Alberty (1956) we represent the combination of enzyme and substrate at the active site as ES and at an adjacent site as SE

$$E + S \underset{k_2}{\overset{k_1}{\rightleftharpoons}} ES \overset{k_3}{\rightarrow} E + P,$$

$$E + S \underset{k_5}{\overset{k_4}{\rightleftharpoons}} SE,$$

$$SE + S \underset{k_7}{\overset{k_6}{\rightleftharpoons}} SES \overset{k_8}{\rightarrow} SE + P.$$

If $k_8 < k_3$ inhibition will result at high substrate concentration while if $k_8 > k_3$ there will be activation. An enzyme whose kinetics can be explained by the above reaction scheme is malic dehydrogenase (Davies and Kun, 1957) which shows substrate

86

activation in the presence of excess malate $(k_8 > k_3)$ and substrate inhibition in the presence of excess oxaloacetate $(k_8 < k_3)$.

Consequently the kinetic treatment employed by Foster, McRae and Bonner cannot as they claim 'constitute a critical experimental test of the two-point attachment concept as applied to auxin action'. It is important to recognise the limitations of kinetic analysis, which cannot be used to prove a particular reaction mechanism, but only to eliminate certain postulated mechanisms.

Michaelis–Menten-type kinetics have been applied to the study of salt uptake and the demonstration that one ion competitively inhibits the uptake of another has been taken to mean that the ions compete for the same site (Epstein and Hagen, 1952). It is true that the observed competition is to be expected if a single site is involved, since each ion must displace the other in order to combine; but it is important to recognise that other possible interactions are also compatible with the observed kinetics of competition. Thus, if the combination of ion and carrier at one site affects the dissociation constant at another site, the kinetics of competitive inhibition will be observed. It is also worth noting that equations fitting a Michaelis–Menten-type curve can be derived without assuming a complex between ion and carrier (Medwedew, 1937). Before applying enzyme kinetics to a physiological problem, it is worth noting the conclusion drawn by Ogston (1955) that 'no deductions about the manner of combination can be drawn from kinetic studies alone without the use of additional considerations such as chemical reactivities, direct demonstration of chemical change and comparisons of stereochemical structures'.

The competition between two compounds for a single site has produced some misunderstanding. Thus Stafford (1956) observed that plant extracts can oxidise tartronate and malate and in order to determine whether or not separate enzymes were involved, she diluted the extract until activity with tartronate could no longer be detected. She then found that tartronate was a competitive inhibitor of malate oxidation and concluded that this finding 'would argue against the activity being due to malic dehydrogenase'. It is difficult to accept this argument since Stafford's results are readily explained on the basis of a single enzyme. Thorn (1949) and Whittaker and Adams (1949) have independently shown that if two substrates can combine at the same enzyme site, the total initial velocity vt is given by

$$v_t = \frac{V_{1\,\mathrm{max}}S_1}{S_1 + K_{m_1}}\left(1 + \frac{S_2}{K_{m_2}}\right) + \frac{V_{2\,\mathrm{max}}S_2}{S_2 + K_{m_2}}\left(1 + \frac{S_1}{K_{m_1}}\right),$$

(where V_{max} represents the maximum velocity in the presence of a saturating concentration of substrate, K_m is the Michaelis constant and S the concentration of each substrate.

If we call malate S_1 and tartronate S_2 we note that under the conditions employed by Stafford (enzyme too dilute to show activity with S_2) the right-hand term in the above equation is zero and the equation reduces to the simple equation for competitive inhibition.

Returning to the topic of the link between carbohydrate and amino acid metabolism, it appears probable that the most important link is formed by the couple α-ketoglutarate and glutamate. The biosynthesis of secondary amino acids by trans-amination with glutamate does not involve a change in the sum of glutamate and α-oxoglutarate, but, as can be seen from the following equations, a net synthesis of α-amino acids requires a high ratio of DPNH/DPN:

$$NH_4^+ + DPNH + \alpha\text{-oxoglutarate}$$
$$\rightleftharpoons glutamate + DPN^+ + H_2O,$$

$$\alpha\text{-ketoacid} + glutamate$$
$$\rightleftharpoons \alpha\text{-amino acid} + \alpha\text{-oxoglutarate}.$$

Sum $NH_4^+ + DPNH + 2\text{-ketoacid}$
$$\rightleftharpoons \alpha\text{-amino acid} + DPN^+ + H_2O.$$

Under aerobic conditions there will be competition for DPNH between glutamine dehydrogenase and the mito-chondrial DPNH oxidase system with its associated phosphoryla-tion of ADP. To this extent the requirements for protein synthesis (ATP and amino acids) are opposed and there may be a delicate balance between oxidative phosphorylation and amino acid synthesis.

In addition to acting as a source of α-amino groups, glutamate is the precursor of a number of other amino acids (fig. 30). The synthesis of amino acids from glutamate and aspartate (fig. 31) involves the removal of their ketoacid analogues from the reactions of the Krebs cycle. As pointed out previously (p. 13), the dual function of the Krebs cycle as a catalytic respiratory system and a biosynthetic pathway, requires that the organic acids utilised in biosynthesis be replenished, either by carbon dioxide fixation or by a $C_2 + C_2$ condensation. Respiration, carbon dioxide fixation and protein synthesis are thus closely

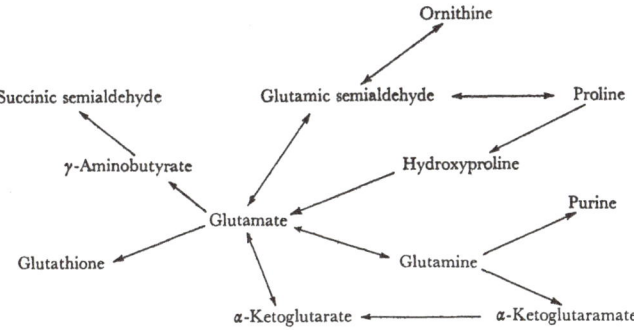

Fig. 30. Metabolic reactions involving glutamate. α-Ketoglutarate may be formed from glutamate by oxidative deamination (glutamic dehydrogenase) or by trans-amination with a number of ketoacids. The decarboxylation of glutamate to give γ-aminobutyrate is shown as an irreversible reaction, but the possibility of reversal during photosynthesis has been considered by Warburg's school. Hydroxyproline, a constituent of collagen, is also a constituent of a protein associated with randomly proliferating plant cells.

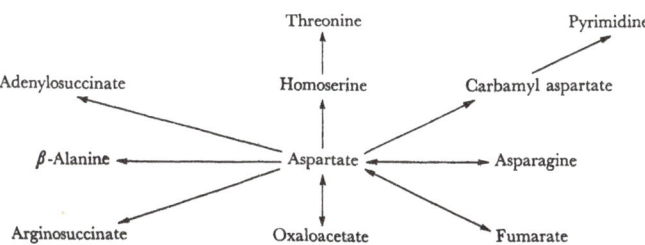

Fig. 31. Metabolic reactions involving aspartate. The interconversion of aspartate and fumarate is well established in bacteria, but is less certain in plants. Decarb-oxylation of aspartate to give β-alanine has not been demonstrated in plants, though it is secreted by pea-root nodules. Pea roots convert aspartate to homoserine which is transported to other parts of the plant. Many other plants accumulate threonine.

interrelated and the complexity of their interactions compli-cates the interpretation of experiments designed to test the hypothesis of a protein cycle.

In recent years increasing attention has been paid to the metabolism of one- and two-carbon compounds, so that in 1957 the editors of the *Annual Review of Biochemistry* were able to write 'one and two carbon biochemistry has interpenetrated the entire metabolic framework'. Acetyl-CoA is a key metabolite (fig. 32) providing the point of entry into the Krebs cycle for fatty acids, carbohydrates and certain amino acids. In addition it is the building unit for the biosynthesis of fats, terpenoid substances such as carotene, cholesterol and rubber and as a participant in the glyoxalate cycle is implicated in the formation of carbohydrates from fats.

The glyoxalate cycle (fig. 27) may be considered as a mechanism for the oxidation of acetate to glyoxalate. It will be noted that the oxidation of acetate to glyoxalate represents the sum of the reactions of the cycle and that the carbon atoms of acetate do not, in a single turn of the cycle, become those of glyoxalate. Thus ^{14}C-labelled acetate entering the glyoxalate cycle would

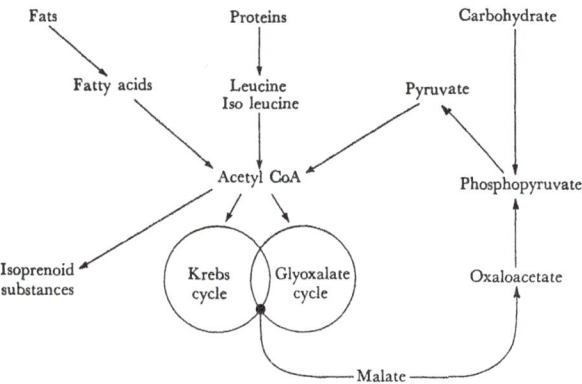

Fig. 32. Metabolic reactions involving acetyl-CoA.

initially label the C-4 acids, but not glyoxalate. The distribution of the key enzyme *iso*citritase seems to be limited to those tissues in which active fat metabolism is taking place (Carpenter and Beevers, 1958) and presumably the glyoxalate cycle is similarly limited.

The synthesis of isoprenoid compounds from acetate is now well established. Important advances in understanding the biosynthetic pathway were the identification of mevalonic acid (3-methyl-3-hydroxyvalerolactone) as the acetate-replacing factor in *Lactobacillus acidophilus*, and the observation that mevalonic acid is efficiently converted to cholesterol by cell-free extracts of liver (Wolf, Hoffman, Aldrich, Skeggs, Wright

90

and Folkers (1956); Tavormina, Gibbs and Huff (1956)). The incorporation of ^{14}C-labelled acetate and mevalonate into β-carotene has now been demonstrated in cell-free extracts of carrot (Braithwaite and Goodwin, 1958) and it seems probable that the scheme outlined in fig. 33 is essentially correct.

The first detectable two-carbon compound to be labelled during photosynthesis in $^{14}CO_2$ is glycolate (Schou, Benson,

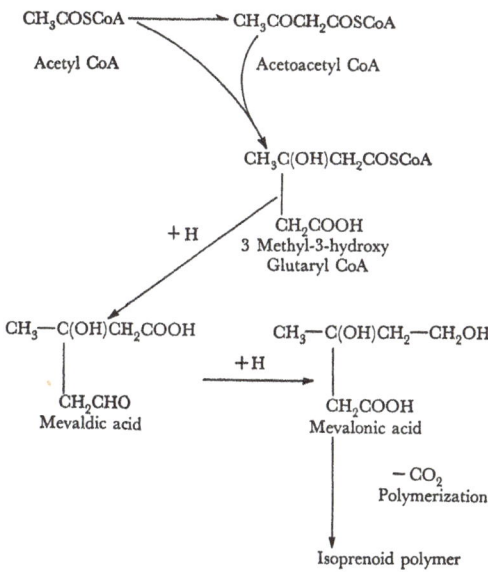

Fig. 33. Scheme for the biosynthesis of isoprenoid compounds.

Bassham and Calvin, 1950), but the mechanism of its biosynthesis is uncertain. Weissbach and Horecker (1955) have given evidence that ribose-5-phosphate-1-^{14}C yields glycine-2-^{14}C in cell-free preparations obtained from spinach leaves. They suggested that glycolaldehyde was an intermediate in the synthesis of glycine from ribose-5-phosphate and postulated the release of glycolaldehyde from the compound between transketolase and 'active' glycolaldehyde. An alternative scheme (fig. 34) for the biosynthesis of glycolaldehyde from phosphoglycerate has been discussed (Davies, Hanford and Wilkinson, 1959), but the relative importance of either pathway is unknown. An aldehyde dehydrogenase active with glycolaldehyde has been demonstrated in pea epicotyls (Davies *et al.* 1959) and in spinach leaves. The pea enzyme has a high affinity for glycolaldehyde, but the enzyme will also oxidise acetaldehyde to acetate:

$$CH_2OHCHO + DPN^+ \underset{}{\overset{H_2O}{\rightleftharpoons}} CH_2OHCOOH + DPNH + H^+$$

Glycolic acid is a substrate for the flavoprotein glycolic acid oxidase which is probably present in all green leaves, and has some properties suggesting the adaptive nature of the enzyme.

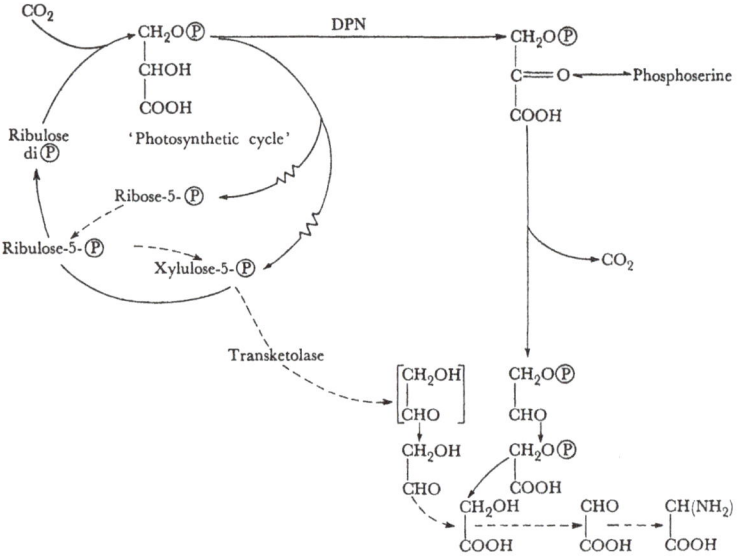

Fig. 34. Scheme for the biosynthesis of glycine. 'Active' glycolaldehyde is shown in brackets as a product of transketolase. It may also be formed by the action of aldolase on xylulose-1-phosphate.

Thus the appearance of glycolic acid oxidase in aetiolated plants can be induced by illumination or by spraying with glycolate (Tolbert and Cohan, 1953):

$$\begin{array}{c} CH_2OH \\ | \\ COOH \end{array} \xrightarrow{+O_2} \begin{array}{c} CHO \\ | \\ COOH \end{array} + H_2O_2$$

The product of the reaction, glyoxalate, is readily transaminated to yield glycine, but the results of feeding experiments indicate that glycolate is also a precursor of the hydroxymethyl group of serine. When wheat leaves were fed with uniformly labelled glycolate, the serine produced was found to be uniformly labelled (Tolbert and Cohan, 1953). Assuming that glycolate is the precursor of glycine and that serine is formed from glycine, it follows that the β-carbon of serine must have come from glycolate without dilution. Similarly it was found that when glycolaldehyde-2-^{14}C and glycine-2-^{14}C were fed to turnip slices, the methyl groups of S-methyl cysteine and methionine became labelled (Davies and Wilkinson, 1959). It appears probable that in the light, most methyl groups are derived from the α-carbon of glycolaldehyde and in darkness from 3-phosphoglycerate, via the β-carbon of serine (Hanford and Davies, 1958).

The formation of a C-1 unit from glyoxalate may be due to a non-enzymic reaction with hydrogen peroxide resulting in the formation of formic acid (Kenten and Mann, 1952). If such is the case, the formate must be reduced to the level of formaldehyde before the C_1 unit can become the hydroxymethyl carbon of serine. The reduction of formate to formaldehyde necessitates the activation of formate and spectrophotometric and chromatographic evidence indicates that in turnips the reactions involved are similar to those demonstrated in animals and bacteria:

$$\text{Formate} + \text{ATP} + \text{THFA} \rightleftharpoons \text{ADP} + \text{P}_1 + \text{formyl THFA},$$
$$\text{Formyl THFA} + \text{TPNH} + \text{H}^+$$
$$\rightleftharpoons \text{TPN}^+ + \text{hydroxymethyl THFA}.$$

In these equations THFA represents tetrahydrofolic acid:

5:6:7:8-tetrahydrofolic acid

This coenzyme may be called the coenzyme of C-1 metabolism and some of the reactions in which it participates are shown in fig. 34. The names formyl-THFA and hydroxymethyl-THFA are used to indicate the oxidation level of the intermediate and not the exact chemical nature. Thus three hydroxy-methyl derivatives of THFA, the ^{5}N and ^{10}N isomers and the ^{5}N^{10}N bridge compound are all interconvertible in the presence of the appropriate enzymes, and the same is true for the formyl derivatives. An excellent account of the chemistry of the folic acid coenzymes has been provided by Huennekens and Osborn (1959).

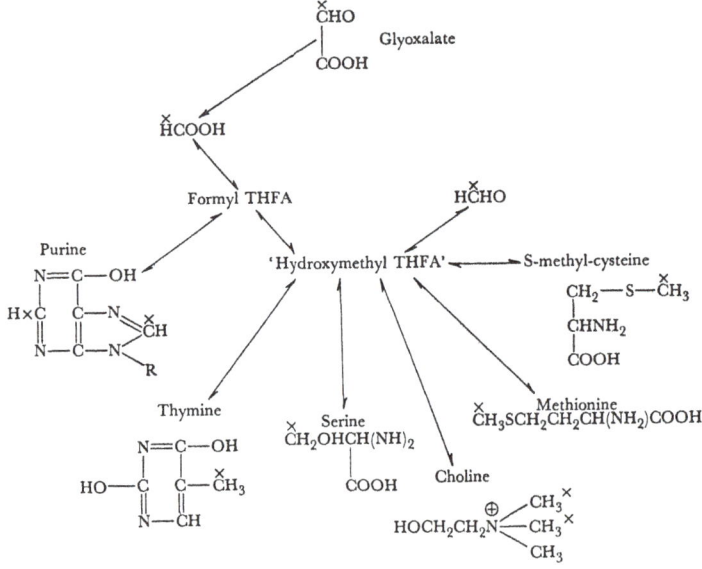

Fig. 35. Metabolic reaction of C-1 compounds. The carbon atoms of the various compounds involved in C-1 metabolism are indicated by ×.

There is much evidence from feeding experiments that formate and formaldehyde can label the compounds shown in fig. 35, but the only direct evidence for the participation of folic acid coenzymes in plant metabolism comes from the demonstration that THFA is necessary for the activity of serine-aldolase isolated from turnips (Wilkinson and Davies, 1958). The mechanism whereby hydroxymethyl groups are reduced to methyl groups is uncertain. It is possible that only one such reduction occurs with the formation of methionine which can then donate its methyl group, perhaps after activation to S-adenosyl methionine, to other compounds. Thus Sato, Byerrum, Albersheim and Bonner (1958) have shown that methionine is converted in plants to methionine sulphoxide and the methyl

94

group is transferred to pectin—a reaction which may play an important part in cell elongation. However, experiments in the author's laboratory have shown the rapid interconversion of the methyl groups of methionine and methylcysteine with the hydroxymethyl group of serine. The rapidity of these reversible reactions makes the interpretation of feeding experiments difficult and identity of a methyl donor can only be established when the individual enzymes have been isolated.

CHAPTER 7

Conclusion

EVOLUTIONARY theory leads to the expectation that the pattern of intermediary metabolism is similar in animals, plants and bacteria. Plants, however, synthesise all their organic constituents, whereas animals are unable to synthesise a number of compounds, which they obtain in their food. Consequently part of the pattern of intermediary metabolism is absent from animals and heterotrophic bacteria. Careful investigation has shown that not only patterns, but individual reactions and the enzymes catalysing them are frequently similar in plants and animals. For example, the reactions of the Krebs cycle are identical in plants and animals and the similarity extends to the properties of the individual enzymes. Thus malic dehydrogenase from heart muscle is able to oxidise malate, meso- and $D(-)$-tartrate, dihydroxyfumarate and tartronate; the enzyme from peas and cauliflowers has the same specificity.

In the absence of information about a particular aspect of plant metabolism, it may be tempting to assume that the general similarity of plant and animal metabolism is applicable in a special case. There are, however, sufficient exceptions to make the assumption dangerous. For example, at the enzyme level, animal tissues possess a glutamic dehydrogenase which can function with DPN or TPN, the plant enzyme is specific for DPN. At the reaction level, guanosinetriphosphate (GTP) is a reactant in the following reactions catalysed by enzymes from animals:

Succinyl-CoA + Pi + CO_2 + GDP
$$\rightleftharpoons \text{succinate} + \text{CoASH} + \text{GTP},$$

Phosphopyruvate + GDP + CO_2 $\rightleftharpoons$ oxaloacetate + GTP,

whereas the corresponding reactions in plants involve ATP in place of GTP.

Differences are also to be found at the level of metabolic

patterns and it is conceivable that in some cases they may reflect phylogenetic differences. Bacteria and *Chlorella* make lysine from aspartate via diaminopimelic acid, whereas fungi and *Euglena* make lysine via aminoadipic acid (Vogel, 1959).

The classification of animals and plants is based on structural differences which are readily observed. Plants, unlike animals, produce and accumulate a large variety of compounds making it possible to classify plants on the chemistry of their products. The great variety of secondary plant products raises two questions; their participation in intermediary metabolism and their biological significance.

The diversity of plant products and the restriction of certain compounds to individual species or groups, make it unlikely that they play a major role in intermediary metabolism, but it would be most unwise to group them all as waste products of metabolism. Shikimic acid was once thought to be a rare plant acid, but is now known as an important intermediate in the formation of the aromatic ring. Some of the secondary products may not be as metabolically inert as was once believed. In aetiolated buckwheat seedlings, rutin and cyanidin glycoside were both labelled after feeding with carboxyl-labelled phenylalanine. After reaching a maximum, the specific activity of rutin and cyanidin glycoside declined, suggesting that both substances are metabolised by buckwheat and cannot be regarded as end-products of metabolism (Grisebach and Bopp, 1959).

The large variety of secondary products would seem to argue against the view that specific enzymes are responsible for their synthesis, but the stereospecific formation of many compounds suggests the participation of enzymes. In the synthesis of certain alkaloids it is likely that compounds with two amino groups are oxidised to aminoaldehydes or aminoketones by plant amine-oxidase and that these compounds spontaneously cyclise (Mothes, 1959). Thus the amine oxidase catalysed oxidation of 1:4-diaminobutane and 1:5-diaminopentane leads to the formation of pyrrolidene and piperidine compounds (Mann and Smithies, 1955). The unsaturated ring compounds formed by cyclisation of the products of amine oxidation can condense with and decarboxylate α-ketoacids (Clarke and Mann, 1959) as suggested by Robinson (1917). Some of the non-protein amino acids found in plants may also be formed by relatively unspecific enzyme reactions (Fowden, 1958). Thus purified, glutamic-oxaloacetate-transaminase will transfer the amino group from γ-methylene-glutamate, γ-hydroxyglutamate, hydroxy-aspartate, cysteate and cysteinesulphinate to α-ketoglutarate or oxaloacetate (Ellis, 1959).

The biological significance of secondary plant products can only be discussed in broad terms, because of the diversity of compounds. Some may represent waste products, others the products of a detoxification mechanism. Thus the formation of glucosides of phloroglucinol and other phenols can be observed when leaves are supplied with glucose-^{14}C and a phenol (Hutchinson, Roy and Towers, 1958). Fraenkel (1959) has suggested that secondary plant products arose as a means of protecting plants from insects, but now guide insects to particular host plants.

Finally must be mentioned the effect of secondary plant products on animals. The alkaloids have been studied because they produce readily detected responses, particularly on the nervous system. The toxic principles of many plants have been identified and in the case of fluoracetic acid (the toxic principle of *Dichapetalum cymosum*) it has been shown that fluoracetate gives rise to fluorocitrate (Peters, 1957). In many cases the toxic principle is not metabolised, thus djenkolic acid, which is the toxic principle of the djenkol bean (*Pithecolobium lobatum*), appears to damage the ureters because it appears as needle-shaped crystals in the urine and produces mechanical damage. These and other products have been studied because their effect on animals is easily detected. The metabolism of compounds which have less obvious effects on animals has received scant attention. Some compounds may be beneficial, but it is also conceivable that others have minor, but chronic effects on animal health. It is probable that the world's population will continue to increase for many years, and it may be necessary to extend the variety of plants eaten by man. It will then be necessary to watch for harmful effects and to assess the nutritional value of the new foodstuffs. The nutritional value of secondary plant products has been little studied. For example, cabbages, cauliflowers and turnips contain S methylcysteine; it would be of interest to know if the methyl group can be utilised by animals and so replace methionine in the diet. Co-operation between plant and animal biochemists is important since there are relatively few plant biochemists and much time is spent demonstrating the presence in plants of enzymes and reactions which have been demonstrated in animals and bacteria. Considerable advances could be expected if investigators who have developed an assay for an enzyme or metabolic pathway would apply their method to a limited range of plant material.

REFERENCES

ALBERTY, R. A. (1956). *Advanc. Enzymol.* **17**, 1.
ALTERMATT, H. A. and NEISH, A. C. (1956). *Canad. J. Biochem. Physiol.* **34**, 405.
ARNON, D. I., ROSENBERG, L. L. and WHATLEY, F. R. (1954). *Nature, Lond.*, **173**, 1132.
ARNON, D. I., WHATLEY, F. R. and ALLEN, M. B. (1958). *Science,* **127**, 1026.
ARONOFF, S. (1956). *Techniques of Radiobiochemistry.* Iowa State College Press.
ARONOFF, S. (1957). *Bot. Rev.* **23**, 65.
AVRON, M. and BIALE, J. B. (1957). *J. Biol. Chem.* **225**, 699.
BADDELEY, J. and BUCHANAN, J. G. (1958). *Quart. Rev. Chem. Soc., Lond.*, **12**, 152.
BALL, E. G. (1938). *Biochem. Z.* **295**, 262.
BANDURSKI, R. S. and TEAS, H. J. (1957). *Plant Physiol.* **32**, 643.
BARKER, J. and MAPSON, L. W. (1955). *Proc. Roy. Soc. B,* **141**, 321.
BASSHAM, J. A. and CALVIN, M. (1957). *Path of Carbon in Photosynthesis.* Englewood Cliffs, N.J.: Prentice-Hall Inc.
BASSHAM, J. A., SHIBATA, K., STEENBERG, K., BOURDON, J. and CALVIN, M. (1956). *Univ. Calif. Rad. Lab.* p. 3331.
BECK, W. G. (1958). *J. Biol. Chem.* **232**, 251.
BEEVERS, H. (1957). *Biochem. J.* **66**, 23P.
BEINERT, H. and CRANE, F. L. (1956). In *Inorganic Nitrogen Metabolism,* ed. W. D. McElroy and B. Glass. Baltimore: Johns Hopkins Press.
BEINERT, H. and STEYN-PARVÉ, E. (1957). *Proc. Int. Symp. on Enz. Chem.* p. 326. London: Pergamon Press.
BENNET-CLARK, T. A. and BEXON, D. (1943). *New Phytol.* **42**, 65.
BIDWELL, R. G. S., KROTKOV, G. and REED, G. B. (1955). *Canad. J. Bot.* **55**, 189.
BIRCH, A. J. and DONOVAN, F. W. (1953). *Aust. J. Chem.* **6**, 360.
BLACKMAN, F. F. (1905). *Ann. Bot.* **19**, 281.
BLACKMAN, F. F. and PARIJA, P. (1928). *Proc. Roy. Soc. B,* **103**, 412.
BOCK, R. M. and LING (1954). *Anal. Chem.* **26**, 1543.
BOGORAD, L. (1958). *Annu. Rev. Plant Physiol.* **9**, 417.
BONNER, J. (1949). *Amer. J. Bot.* **36**, 429.
BONNER, J. (1950). *Plant Biochemistry.* New York: Academic Press.
BONNER, J. (1959). *Amer. J. Bot.* **46**, 58.
BONNER, J. and BANDURSKI, R. S. (1952). *Annu. Rev. Plant Physiol.* **3**, 59.
BRADY, R. O. (1958). *Proc. Nat. Acad. Sci., Wash.*, **44**, 993.
BRAITHWAITE, G. D. and GOODWIN, T. W. (1958). *Nature, Lond.*, **182**, 1304.
BRIGGS, G. E. and HALDANE, J. B. S. (1925). *Biochem. J.* **19**, 338.
BROWN, A. H. (1953). *Amer. J. Bot.* **40**, 719.
BRUMMOND, D. O. and BURRIS, R. H. (1954). *J. Biol. Chem.* **209**, 755.

BUCHANAN, J. M. and STANDISH, C. H. (1959). *Advanc. Enzymol.* **21**, 199.

BURK, D. (1939). *Cold Spr. Harb. Symp. Quant. Biol.* **7**, 420.

CALVIN, M. and BARLTROP, J. A. (1952). *J. Amer. Chem. Soc.* **74**, 6153.

CAPUTTO, R., LELOIR, L. F., TRUCCO, R. E., CARDINI, C. E. and PALADINI, A. (1949). *J. Biol. Chem.* **179**, 497.

CARPENTER, W. and BEEVERS, H. (1958). *Abstr. Amer. Soc. Plant Physiol.* **33**, Suppl. xxxv.

CHIBNALL, A. C. (1939). *Protein Metabolism in the Plant.* Yale Univ. Press.

CLARKE, A. J. and MANN, P. J. G. (1959). *Biochem. J.* **71**, 596.

CLAUDE, A. (1940). *Science*, **91**, 77.

COHN, M. (1949). *J. Biol. Chem.* **180**, 771.

COHN, M. (1957). *Bact. Rev.* **21**, 140.

COHN, M. and DRYSDALE, A. C. (1955). *J. Biol. Chem.* **216**, 831.

COMMONER, B. and THIMANN, K. V. (1941). *J. Gen. Physiol.* **24**, 279.

COOIL, B. J. (1952). *Plant Physiol.* **27**, 49.

CORI, C. F., VELICK, S. F. and CORI, G. T. (1950). *Biochem. biophys. acta*, **4**, 160.

CRANE, F. G. and LESTER, R. C. (1958). *Plant Physiol.* **33**, Suppl. vii.

CROMWELL, B. T. (1944). *Biochem. J.* **37**, 722.

CROZIER, W. J. (1924). *J. Gen. Physiol.* **7**, 180.

DAVIES, D. D. (1953). *J. Exp. Bot.* **4**, 177.

DAVIES, D. D. (1956). *J. Exp. Bot.* **7**, 203.

DAVIES, D. D. (1959). *Biol. Rev.* **34**, 407.

DAVIES, D. D. and ELLIS, R. J. (1959). In preparation.

DAVIES, D. D. and KUN, E. (1957). *Biochem. J.* **66**, 307.

DAVIES, D. D. and WILKINSON, A. P. (1959). Unpublished results.

DAVIES, D. D., HANFORD, J. and WILKINSON, A. P. (1959). *Symp. Soc. Exp. Biol.* **13**, 353.

DE LA HABA, G., LEDER, I. and RACKER, E. (1955). *J. Biol. Chem.* **214**, 409.

DICKENS, F. (1938). *Biochem. J.* **32**, 1626 and 1645.

DIXON, M. (1937). *Biol. Rev.* **12**, 431.

DIXON, M. (1949). *Multi-Enzyme Systems.* Cambridge Univ. Press.

DIXON, M. and WEBB, E. C. (1958). *Enzymes.* London: Longmans, Green, and Co. Ltd.

DOONER, R. W., KAHN, A. and WILDMAN, S. G. (1957). *J. Biol. Chem.* **229**, 945.

EDELMAN, J. (1956). *Advanc. Enzymol.* **17**, 189.

EDELMAN, J. (1958). Private communication.

ELLIS, R. J. (1959). Ph.D. Thesis. Univ. of London.

EMBDEN, G. and JOST, H. (1934). *Z. Phys. Chem.* **230**, 69.

EPSTEIN, E. and HAGEN, C. E. (1952). *Plant. Physiol.* **27**, 457.

FAWCETT, C. H., INGRAM, J. M. A. and WAIN, R. L. (1954). *Proc. Roy. Soc.* B, **142**, 525.

FISCHER, H. O. L. and BAER, E. (1932). *Ber.* **65**, 337.

FOLKES, B. F. (1959). *Symp. Soc. Exp. Biol.* **13**, 126.

FOLKES, B. F. and YEMM, E. W. (1958). *New Phytol.* **57**, 106.

FOSTER, J. W. (1949). *Chemical Activities of Fungi.* New York: Academic Press.

FOSTER, R. J., McRAE, D. H. and BONNER, J. (1952). *Proc. Nat. Acad. Sci., Wash.*, **38**, 1014.

FOWDEN, L. (1958). *Biol. Rev.* **33**, 393.
FRAENKEL, G. S. (1959). *Science*, **129**, 1466.
FRUTON, J. S. and SIMMONDS, S. (1953). *General Biochemistry*. New York: John Wiley and Sons.
GIBBS, M. (1951). *Plant Physiol.* **26**, 549.
GIBBS, M. and BEEVERS, H. (1955). *Plant Physiol.* **30**, 343.
GILLESPIE, R. J., MAW, G. A. and VERNON, C. A. (1953). *Nature, Lond.*, **171**, 1147.
GINSBERG, V., STUMPF, P. K. and HASSID, W. Z. (1956). *J. Biol. Chem.* **223**, 977.
GRABE, B. (1958). *Biochem. biophys. acta*, **30**, 560.
GRACE, N. H. (1939). *Canad. J. Res.* **17**, 247.
GREEN, D. E. (1954). *Biol. Rev.* **29**, 330.
GREEN, D. E. (1957). *Symp. Soc. Exp. Biol.* **10**, 30.
GREGORY, F. G. and SEN, P. J. (1939). *Ann. Bot.* (N.S.), **1**, 521.
GRISEBACH, H. and BOPP, M. (1959). *Z. Naturf.* **14B**, 485.
HACKETT, D. P. (1955). *Int. Rev. Cytol.* **4**, 143.
HACKETT, D. P. and THIMANN, K. V. (1953). *Amer. J. Bot.* **40**, 183.
HANES, C. S. (1940). *Proc. Roy. Soc. B*, **128**, 421.
HANFORD, J. and DAVIES, D. D. (1958). *Nature, Lond.*, **182**, 532.
HARRISON, K. (1958). *Nature, Lond.*, **181**, 1131.
HARTMAN, S. C. and BUCHANAN, J. M. (1959). *Ann. Rev. Biochem.* **28**, 365.
HEARON, J. Z. (1952). *Physiol. Rev.* **32**, 499.
HEIDELBERGER, C. and POTTER, V. R. (1951). In Ciba Foundation Symposium, *Use of Isotopes in Biology and Medicine*.
HEITZ, E. (1957). *Z. Naturf.* **12B**, 283 and 516.
HEVESEY, G., LINDERSTROM-LANG, K., KESTEN, A. S. and CARSTEN, O. (1940). *C.R. Lab. Carlsberg*, **23**, 213.
HIATT, H. H., GOLDSTEIN, M., LAREAU, J. and HORECKER, B. L. (1958). *J. Biol. Chem.* **231**, 303.
HILL, R. (1954). *Nature, Lond.*, **174**, 501.
HILL, R. and SCARISBRICK, R. (1940). *Nature, Lond.*, **146**, 61; and *Proc. Roy. Soc. B*, **129**, 238.
HILL, R. and WHITTINGHAM, C. P. (1953). *Photosynthesis*. London: Methuen and Co.
HOBERMAN, H. D. (1958). *J. Biol. Chem.* **233**, 1045.
HOPKINS, G. (1913). *Report of the 83rd Meeting of the British Association*, p. 652.
HOPKINS, G. (1932). *Proc. Roy. Soc. B*, **112**, 159.
HORECKER, B. L., SMYRNIOTIS, P. Z. and HURWITZ, J. (1956). *J. Biol. Chem.* **223**, 1009.
HUENNEKENS, F. M. and OSBORN, M. J. (1959). *Advanc. Enzymol.* **21**, 369.
HUTCHINSON, A., ROY, C. and TOWERS, G. H. N. (1958). *Nature, Lond.*, **181**, 841.
JAMES, W. O. (1957). *Advanc. Enzymol.* **18**, 281.
JAMES, W. O. and DAS, V. S. R. (1957). *New Phytol.* **56**, 325.
JAMES, W. O., JAMES, G. M. and BUNTING, A. H. (1941). *Biochem. J.* **35**, 588.
JAMES, W. O. and SLATER, W. G. (1959). *Proc. Roy. Soc. B*, **150**, 192.
KENTEN, R. H. and MANN, P. J. G. (1952). *Biochem. J.* **52**, 130.
KNOOP, F. (1904). *Beitr. chem. Physiol. Path.* **6**, 159.

KORKES, S. (1952). In *Phosphorus Metabolism*, II, p. 618, ed. W. D. McElroy and B. Glass. Baltimore: Johns Hopkins Press.

KORNBERG, H. L. and BEEVERS, H. (1957). *Biochim. biophys. acta*, **26**, 531.

KORNBERG, H. L. and KREBS, H. A. (1957). *Nature, Lond.*, **179**, 988.

KREBS, H. A. (1941). *Nature, Lond.*, **147**, 560.

KREBS, H. A. (1947). *Enzymologia*, **12**, 88.

KREBS, H. A. (1953). *Brit. Med. Bull.* **9**, 97.

KREBS, H. A. (1954). *Johns Hopk. Hosp. Bull.* **95**, 19.

KREBS, H. A. (1957). *Endeavour*, **16**, 125.

KREBS, H. A. and KORNBERG, H. L. (1957). *Ergebn. Physiol. biol. Chem. exp. Pharmakol.* **49**, 212.

KRETOVITCH, W. L. (1958). *Advanc. Enzymol.* **20**, 319.

KRIMSKY, I. (1959). *J. Biol. Chem.* **234**, 232.

LATIES, G. C. (1953). *Plant Physiol.* **28**, 557.

LELOIR, L. F. and CARDINI, C. E. (1957). *J. Amer. Chem. Soc.* **79**, 6340.

LELOIR, L. F., CARDINI, C. E. and CHIRIBOGA, J. (1955). *J. Biol. Chem.* **214**, 149.

LEVY, R. H. and VENNESLAND, B. (1957). *J. Biol. Chem.* **228**, 85.

LIPMANN, F. (1933). *Biochem. Z.* **265**, 133.

LIPMANN, F. (1941). *Advanc. Enzymol.* **1**, 99.

LIPMANN, F. (1942). *Symposium on Respiratory Enzymes*, p. 48. Univ. of Wisconsin Press.

LYNEN, F. (1954). *Harvey Lect.* **48**, 210.

LYNEN, F. and KOENIGSBERGER, R. (1951). *Leibigs Ann.* **573**, 60.

MACLACHLAN, G. A. and PORTER, H. K. (1959). *Proc. Roy. Soc.* B, **150**, 460.

MANN, P. J. G. and SMITHIES, W. R. (1955). *Biochem. J.* **61**, 89.

MARTIN, E. M. and MORTON, R. K. (1956). *Biochem. J.* **62**, 696.

MEDWEDEW, G. (1937). *Enzymologia*, **2**, 1, 31, 53.

MILLERD, A. (1953). *Arch. Biochem.* **42**, 149.

MILLERD, A., BONNER, J., AXELROD, B. and BANDURSKI, R. S. (1951). *Proc. Nat. Acad. Sci., Wash.*, **37**, 855.

MOTHES, K. (1959). *Symp. Soc. Exp. Biol.* **13**, 258.

NELSON, C. D. and KROTKOV, G. (1956). *Canad. J. Bot.* **34**, 423.

NEUFELD, E. F., FENGOLD, D. S. and HASSID, W. Z. (1958). *J. Amer. Chem. Soc.* **80**, 4430.

NEWBURGH, R. W. and CHELDELIN, V. H. (1956). *J. Biol. Chem.* **218**, 89.

NYGAARD, A. P. and RUTTER, W. J. (1956). *Acta chem. scand.* **10**, 37.

OESPER, P. (1950). *Arch. Biochem.* **27**, 255.

OGSTON, A. G. (1948). *Nature, Lond.*, **162**, 963.

OGSTON, A. G. (1955). *Disc. Faraday Soc.* **20**, 161.

OPARIN, A. I. and KURSSANOV (1931). *Biochem. Z.* **239**, 1.

PARDEE, A. B. and POTTER, V. R. (1948). *J. Biol. Chem.* **176**, 1085.

PARK, R. B. and BONNER, J. (1958). *J. Biol. Chem.* **233**, 340.

PARNAS, J. K. (1943). *Nature, Lond.*, **151**, 577.

PETERS, R. A. (1929). *J. State Med.* **37**, 1.

PETERS, R. A. (1952). *Proc. Roy. Soc.* B, **139**, 143.

PETERS, R. A. (1957). *Advanc. Enzymol.* **18**, 113.

POTTER, V. R. and HEIDELBERGER, C. (1949). *Nature, Lond.*, **164**, 180.

PRICE, C. A. and THIMANN, K. V. (1954). *Plant Physiol.* **29**, 495.
PRIGOGINE, I. (1955). *Introduction to Thermodynamics of Irreversible Processes.* Springfield, Illinois, U.S.A.: C. Thomas.
RACKER, E. and SCHROEDER, E. A. R. (1958). *Arch. Biochem. Biophys.* **74**, 326.
RALL, T. W. and SUTHERLAND, E. W. (1958). *J. Biol. Chem.* **232**, 1065.
ROBBINS, P. W., TRAUT, R. R. and LIPMANN, F. (1959). *Proc. Nat. Acad. Sci., Wash.,* **45**, 6.
ROBERTS, R. B. (1958). *Microsomal Particles and Protein Synthesis.* Pergamon Press.
ROBINSON, R. (1917). *J. Chem. Soc.* **111**, 876.
ROBINSON, R. (1923). *Annu. Reports Prog. Chem.* p. 100.
ROBINSON, R. (1955). *The Structural Relations of Natural Products.* Clarendon Press.
ROSE, I. A. and RIEDER, S. V. (1955). *J. Amer. Chem. Soc.* **77**, 5764; and *J. Biol. Chem.* **231**, 315.
ROTMAN, B. and SPIEGELMAN, S. (1954). *J. Bact.* **68**, 419.
RUZICKA, L. and STOLL, M. (1922). *Helv. chim. acta,* **5**, 929.
SALTMAN, P. (1953). *J. Biol. Chem.* **200**, 145.
SATO, C. S., BYERRUM, R. V., ALBERSHEIM, P. and BONNER, J. (1958). *J. Biol. Chem.* **233**, 128.
SCHNEIDER, W. C. and HOGEBOOM, G. H. (1951). *Cancer Res.* **11**, 1.
SCHOU, L., BENSON, A. A., BASSHAM, J. A. and CALVIN, M. (1950). *Physiol. Plant.* **3**, 487.
SEEGMILLER, C. G., AXELROD, B. and McGREADY, R. M. (1955). *J. Biol. Chem.* **217**, 765.
SHNEOUR, E. A. and ZABIN, I. (1959). *J. Biol. Chem.* **234**, 770.
SMILLIE, R. M. (1955). *Aust. J. Biol. Sci.* **8**, 186.
SRINIVASAN, P. R. and SPRINSON, D. B. (1959). *J. Biol. Chem.* **234**, 716.
STAFFORD, H. A. (1956). *Plant Physiol.* **31**, 135.
STEKOL, J. A. (1958). *Annu. Rev. Biochem.* **27**, 679.
STEWARD, F. C., BIDWELL, R. G. S. and YEMM, E. W. (1958). *J. Exp. Bot.* **9**, 11.
STEWARD, F. C. and POLLARD, J. K. (1957). *Annu. Rev. Plant Physiol.* **8**, 65.
STROMINGER, J. L. and MAPSON, L. W. (1957). *Biochem. J.* **66**, 567.
STUMPF, P. K. and BARBER, G. A. (1956). *Plant Physiol.* **31**, 304.
STUMPF, P. K. and BARBER, G. A. (1957). *J. Biol. Chem.* **227**, 407.
STUMPF, P. K. and BRADBEER, C. (1959). *Annu. Rev. Plant Physiol.* **10**, 197.
SZENT-GYÖRGYI, A. (1957). *Bioenergetics.* New York: Academic Press.
TANKO, B. (1936). *Biochem. J.* **30**, 692.
TAVORMINA, P. A., GIBBS, M. H. and HUFF, J. M. (1956). *J. Amer. Chem. Soc.* **78**, 4498.
TCHEN, T. and VENNESLAND, B. (1955). *J. Biol. Chem.* **213**, 533.
THORN, M. B. (1949). *Nature, Lond.,* **164**, 27.
TOLBERT, N. E. and COHAN, M. S. (1953). *J. Biol. Chem.* **204**, 649.
TS'O, P. O. P., BONNER, J. and VINOGRAD, J. (1958). *Biochim. biophys. acta,* **30**, 570.
TURNER, J. S. (1951). *Annu. Rev. Plant Physiol.* **2**, 145.
TURNER, J. F. and MAPSON, L. W. (1958). *Nature, Lond.,* **181**, 270.

UMBARGER, A. E. and BROWN, B. (1958). *J. Biol. Chem.* **233**, 415.
VELICK, S. F. (1958). *J. Biol. Chem.* **233**, 1455.
VENNESLAND, B. (1955). *Disc. Faraday Soc.* **20**, 240.
VICKERY, H. B. and PALMER, J. K. (1957). *J. Biol. Chem.* **225**, 629.
VISHNIAC, W., HORECKER, B. L. and OCHOA, S. (1957). *Advanc. Enzymol.* **19**, 1.
VITTORIO, P., KROTKOV, G. and REED, G. B. (1955). *Canad. J. Bot.* **33**, 275.
VOGEL, H. J. (1959). *Biochim. biophys. acta*, **34**, 282.
VON KORFF, R. W. and TWEDT, R. M. (1957). *Biochim. biophys. acta*, **23**, 143.
WAKIL, S. J. (1955). *Biochim. biophys. acta*, **18**, 314.
WAKIL, S. J. (1956). *Biochim. biophys. acta*, **19**, 499.
WALD, G. (1956). In *Enzymes: Units of Biological Structure and Function*, ed. O. H. Goebler. New York: Academic Press.
WEBSTER, G. C. (1959). *Symp. Soc. Exp. Biol.* **13**, 330.
WEISSBACH, A. and HORECKER, B. L. (1955). In *Amino Acid Metabolism*, ed. W. D. McElroy and B. Glass. Baltimore: Johns Hopkins Press.
WHITTAKER, V. P. and ADAMS, D. H. (1949). *Nature, Lond.*, **164**, 314.
WHITTAM, R., BARTLEY, W. and WEBER, G. (1955). *Biochem. J.* **59**, 590.
WILKINSON, A. P. and DAVIES, D. D. (1958). *Nature, Lond.*, **181**, 1070.
WILLIAMS, W. T. (1954). *J. Exp. Bot.* **5**, 343.
WILSON, A. T. and CALVIN, M. (1955). *J. Amer. Chem. Soc.* **77**, 5948.
WOLF, D. E., HOFFMAN, C. H., ALDRICH, P. E., SKEGGS, H. R., WRIGHT, L. and FOLKERS, K. (1956). *J. Amer. Chem. Soc.* **78**, 4499.
WOOD, H. G. and KATZ, J. (1958). *J. Biol. Chem.* **233**, 1279.
WORMSER, E. H. L. and PARDEE, A. B. (1958). *Arch. Biochem. Biophys.* **78**, 416.
YEMM, E. W. and FOLKES, B. F. (1958). *Annu. Rev. Plant Physiol.* **9**, 245.
ZBINOVSKY, V. and BURRIS, R. H. (1952). *Plant Physiol.* **27**, 240.
ZELITCH, I. and OCHOA, S. (1953). *J. Biol. Chem.* **201**, 707.

INDEX

acetate metabolism, 90
acetyl-CoA, 90
adenosine triphosphate
  points of cleavage, 42
  production, 44
alkaloid biosynthesis, 68
amino acid
  activation, 30
  synthesis, 83, 84
amylase, 69
aromatic biosynthesis, 69
aspartate metabolism, 89

Bloom and Stetten ratio, 56

carbohydrate metabolism, 53–8
carbon dioxide
  fixation, 13
  role in fat synthesis, 76
carotene synthesis, 91
citric acid cycle, 13–15
closed systems, 2
coenzyme A, 42, 60
crotonase, 61
cytidine diphosphate compounds, 78
cytochromes, 46

dehydrogenases, 10–12
deuterium studies, 10–12
dinitrophenol, 49
diphosphopyridine nucleotide, 10–12

electron transport, 46
endergonic reaction, 38
entropy, 35
epimerase, 81
equilibrium constants, 35–6
  glycolysis, 71–3
  pentose-phosphate pathway, 21
exergonic reaction, 38

fatty acid
  oxidation, 19, 33, 58–63
  synthesis, 74–6
feed-back
  negative, 15
  positive, 8, 17
fermentation, 7, 43

flavin-adenine dinucleotide, 45–7
folic acid, 94
formate metabolism, 94
free energy, 35

galactowaldenase, 81
glucose-6-phosphate dehydrogenase, 7,
  19, 20, 23, 55, 71
glucose-1-phosphate, 70
glutamate metabolism, 89
glutamic oxaloacetic transaminase,
  88
glyceraldehyde-3-phosphate, 7, 16, 18,
  19, 20, 43
glycine synthesis, 92
glycolic acid, 92
glycolysis, 7, 43, 73
glyoxalate, 74, 90, 92
glyoxalate cycle, 74, 90

hexokinase, 1, 7
hexose-monophosphate shunt, 55
high energy
  bonds, 39
  phosphates, 39–40
Hill reaction, 48
hippuric acid, 59

irreversible reactions, 7
isoalloxazine ring, 47
isocitritase, 74, 90
isoleucine synthesis, 16
isoprene
  biosynthesis, 91
  rules, 67

kinases, 42
K_m (Michaelis-Menten constant), 85–8
Krebs cycle, 13–15, 63, 65

malate synthetase, 75
malic dehydrogenase, 86–8
metalloflavoproteins, 31–2, 60
methyl groups, 94
mevalonate, mevalonic acid, 90–1
Michaelis-Menten kinetics, 85–8
microsomes, 28
mitochondria, 31, 44, 45